P9-DNN-326

FROM DUST TO LIFE

FROM DUST TO LIFE

THE ORIGIN AND EVOLUTION OF OUR SOLAR SYSTEM

JOHN CHAMBERS AND JACQUELINE MITTON

PRINCETON UNIVERSITY PRESS
PRINCETON AND OXFORD

Requests for permission to reproduce material from this work should be sent to Permissions, Princeton University Press
Published by Princeton University Press, 41 William Street, Princeton, New Jersey 08540
In the United Kingdom:
Princeton University Press, 6 Oxford Street, Woodstock, Oxfordshire OX20 1TW

press.princeton.edu

ISBN 978-0-691-14522-8
Library of Congress Control Number: 2013944898

British Library Cataloging-in-Publication Data is available

This book has been composed in Minion Pro

Printed on acid-free paper

Printed in the United States of America

1 3 5 7 9 10 8 6 4 2

To Lindsey, Cieran, and Ceinwyn,
for helping to make this possible. . . . JEC

To my husband Simon,
whose love of books and enthusiasm
for writing are inspirational. . . . JM

CONTENTS

List of Illustrations xi

Preface xv

ONE **Cosmic Archaeology** 1

A fascination with the past 1

A solar system to explain 3

Real worlds 9

Winding back the clock 12

Putting the pieces together 16

TWO **Discovering the Solar System** 19

Measuring the solar system 19

From wandering gods to geometrical constructions 22

The Sun takes center stage 25

Laws and order 27

Gravity rules 29

The missing planet 31

Asteroids enter the scene 34

Rocks in space 36

Uranus behaving badly 37

Completing the inventory 40

THREE **An Evolving Solar System** 43

A changing world 43

A nebulous idea begins to take shape 44

The nebular hypothesis in trouble 48

A chance encounter? 50

Nebular theory resurrected 54

FOUR **The Question of Timing** 56

Reading the cosmic clock 57

Early estimates: ingenious—but wrong 57

Geology versus physics 58

Radioactivity changes everything 61

Hubble and the age of the universe 63

How radioactive timers work 64
Meteorites hold the key 68
Dating the Sun 71
The age of the universe revisited 73

FIVE **Meteorites** 75
A dramatic entrance 75
Where do meteorites come from? 76
Irons and stones 80
Identifying the parents 83
Lunar and Martian meteorites 86
A rare and precious resource 87
What meteorites can tell us 88

SIX **Cosmic Chemistry** 92
Element 43: first a puzzle then a clue 92
An abundance of elements 94
The first elements 96
Cooking in the stellar furnace 98
Building heavier elements 104
Supernovae 105

SEVEN **A Star Is Born** 108
A child of the Milky Way 108
Where stars are born 110
First steps to a solar system 113
The solar system's birth environment 119
Essential ingredients 121

EIGHT **Nursery for Planets** 123
An excess of infrared 123
Two kinds of disks 125
Inside the solar nebula 129
Getting the dust to stick 131
The influence of gas 134
How to build planetesimals 135
The demise of the disk 137

NINE **Worlds of Rock and Metal** 140
Sisters but not twins 140
The era of planetesimals 141
Planetary embryos take over 144

The final four 147
Earth 148
Mercury 153
Venus 158
Mars 161

TEN **the Making of the Moon** 168
The Moon today 169
What the Moon is made of 170
The Moon's orbit 172
The fission theory 174
The capture hypothesis 175
The coaccretion hypothesis 176
The giant impact hypothesis 177
Encounter with Theia 179
Earth, Moon, and tidal forces 181
Late heavy bombardment 183

ELEVEN **Earth, Cradle of Life** 186
The Hadean era 186
The tree of life 191
The building blocks of life 193
The rise of oxygen 196
A favorable climate 199
Snowball Earth 202
Future habitability 204

TWELVE **Worlds of Gas and Ice** 205
Giants of the solar system 205
Building giants by core accretion 211
The disk instability model 214
Spin and tilt 215
Masters of many moons 217
Formation of regular satellites 219
The origin of irregular satellites 220
Rings 221

THIRTEEN **What Happened to the Asteroid Belt?** 225
The asteroid belt today 225
Ground down by collisions? 226
Emptied by gravity? 229
Asteroid families 231

The missing mantle problem 233
Asteroids revealed as worlds 236

FOURTEEN **The Outermost Solar System** 242
Where do comets come from? 242
Centaurs 246
Looking beyond Neptune 247
The Kuiper belt 248
Sedna 251
The nature of trans-Neptunian objects 252
Where have all the Plutos gone? 256
The Nice model 259

FIFTEEN **Epilogue: Paradigms, Problems, and Predictions** 263
The paradigm: solar system evolution in a nutshell 264
Unsolved puzzles 267
Searching the solar system for answers 268
Other planetary systems 271
Future evolution of the solar system 273

Glossary 277

Sources and Further Reading 291

Index 293

ILLUSTRATIONS

Figure 1.1. The layout of the solar system. 4
Figure 1.2. Comet Elst-Pizzaro. 9
Figure 1.3. One of the images of Mars returned by the Mariner 4 spacecraft in July 1965. 12
Figure 1.4. Jets streaming from the nucleus of Comet Hartley 2. 18
Figure 2.1. The transit of Venus of June 5–6, 2012, as seen from Earth orbit. 20
Figure 2.2. A 16th-century representation of the universe with Earth at the center. 25
Figure 2.3. The solar system, as pictured by Nicolaus Copernicus. 27
Figure 3.1. Drawings of nebulae made by William Parsons, 3rd Earl of Rosse. 49
Figure 3.2. An early photograph of a "spiral nebula"—the Andromeda Galaxy—published in 1893. 52
Figure 4.1. The half-life of a radioactive isotope. 62
Figure 4.2. A schematic isochron. 67
Figure 4.3. The Clair C. Patterson Award medal. 69
Figure 5.1. Kirkwood gaps. 78
Figure 5.2. The Yarkovsky effect. 80
Figure 5.3. An etched and polished section of the Gibeon iron meteorite, showing the structure of interlocking crystals. 81
Figure 5.4. A microscope image of loose chondrules. 82
Figure 5.5. A microscope image of a section through the carbonaceous chondrite NWA 989. 83
Figure 5.6. The lunar meteorite ALHA81005, found in the Allan Hills area of Antarctica in 1981. 88
Figure 6.1. The relative abundance of the chemical elements in the solar system. 96
Figure 6.2. Margaret Burbidge, Geoffrey Burbidge, Willy Fowler, and Fred Hoyle. 100
Figure 6.3. Simplified schematic diagrams of two of the main nuclear processes that take place in stars. 101

Figure 6.4. An image of the region around the supernova SN 1987A taken by the Hubble Space Telescope in 2003. 107
Figure 7.1. An artist's conception of the spiral structure of our Milky Way galaxy showing the current location of the Sun. 110
Figure 7.2. A composite, near-infrared image of the Trapezium star cluster in the Orion Nebula. 112
Figure 7.3. A Hubble Space Telescope image of pillar-like clouds of dense gas and dust in the Eagle Nebula. 114
Figure 7.4. The Herbig Haro object HH47, imaged by the Hubble Space Telescope. 116
Figure 7.5. Thirty "proplyds" (protoplanetary disks) in the Orion Nebula, imaged by the Hubble Space Telescope. 117
Figure 7.6. The main stages in the evolution of the Sun from its formation to becoming a white dwarf. 118
Figure 8.1. Infrared excess. 124
Figure 8.2. The dust disk surrounding the star Beta Pictoris, imaged at the European Southern Observatory, with the possible orbit of the star's known planet superimposed. 126
Figure 8.3. Debris disks around two stars, imaged by the Hubble Space Telescope. 127
Figure 9.1. Gravitational focusing. 144
Figure 9.2. Oligarchic growth. 145
Figure 9.3. The interior structure of Earth. 149
Figure 9.4. A near global view of Mercury as seen by the Messenger spacecraft on its first flyby. 157
Figure 9.5. "Pancake" domes on Venus. 159
Figure 9.6. The Mangala Valles region on Mars. 166
Figure 10.1. The contrast between mare and highland areas on the Moon. 169
Figure 10.2. A reflector set up on the surface of the Moon by Apollo 14 astronauts. 173
Figure 10.3. A simulation of how the Moon may have formed following a giant impact on Earth. 179
Figure 11.1. A Hadean zircon, magnified 200 times and viewed in transmitted light. 190
Figure 11.2. A simplified version of the "tree of life." 192
Figure 11.3. A timeline for the development of life on Earth, showing the four main eras since Earth formed. 197
Figure 11.4. The carbon-silicon cycle—how plate tectonics contributes to stabilizing Earth's climate. 201

Figure 12.1. Jupiter, imaged by the Hubble Space Telescope in 2009. 206
Figure 12.2. The interior structure of the giant planets. 208
Figure 12.3. Uranus and Neptune. 209
Figure 12.4. Jupiter's Galilean moons to scale, imaged by the New Horizons spacecraft. 218
Figure 12.5. Detailed structure in Saturn's B ring imaged by the Cassini spacecraft in 2009. 223
Figure 13.1. The impact basin Rheasilvia in Vesta's southern hemisphere. 228
Figure 13.2. Ida and Dactyl. 237
Figure 13.3. The near-Earth asteroid Eros. 238
Figure 13.4. Close-ups of Eros taken by the NEAR Shoemaker spacecraft during low-altitude passes in 2001. 239
Figure 13.5. The main-belt asteroid Lutetia. 240
Figure 14.1. A composite image of Comet Wild 2. 245
Figure 14.2. Maps of Pluto's surface. 254
Figure 14.3. A dust particle collected by the Stardust spacecraft during its mission to Comet Wild 2. 255
Figure 14.4. A mosaic of Neptune's moon Triton taken in 1989 by Voyager 2 during its flyby of the Neptune system. 257
Figure 15.1. A timeline of events in the early evolution of the solar system. 266
Figure 15.2. A comparison between the inner solar system and the Kepler-47 system. 273

Tables

Table 1.1. Selected properties of the major planets 6
Table 5.1. The basic classification of meteorites 86
Table 9.1. Physical properties of the terrestrial planets 148
Table 12.1. Mass, diameter, and density of the giant planets 209
Table 12.2. Obliquity and rotation periods of the giant planets 216
Table 13.1. Properties of selected asteroids 227
Table 14.1. Properties of selected trans-Neptunian objects 249

EVERYONE IS FASCINATED BY ORIGINS. We all want to know where we come from, what life was like in the past, how we fit into the larger scheme of things. Each generation has tried to answer these questions in its own way, from mythological creation stories told around the campfire to detailed accounts in religious and philosophical texts. In the past few centuries, the rise of the scientific method has provided a new way to think about these age-old questions, offering the possibility of definitive answers for the first time.

The origin, evolution, and nature of our solar system—the Sun and its family of planets, moons, comets, and asteroids—form a central mystery in the story of our beginnings. Many features of the solar system are essential for sustaining life today, including the longevity and stability of the Sun itself, the existence of water, carbon, nitrogen, and other essential life-building materials, and the size and orbit of Earth, which ensure that our climate is just right for life to flourish and has been so for billions of years. Even the other planets may play a role, helping to supply the raw materials of life to the early Earth, and keeping potentially dangerous impactors at a safe distance. Astronomers have recently found hundreds of other planetary systems in the universe, one of the great discoveries of our time, yet it remains to be seen whether truly Earth-like planets are commonplace or whether a unique series of events during the solar system's formation gave rise to our very special world.

A fascination with the past and the role played by the solar system in determining who we are today provided the motivation for writing this book. Our goal is to tell the story of how the solar system began, as scientists see it, and to describe some key events in its history. At the same time, we look at how scientists have come to appreciate the solar system in all its wondrous detail, how they have slowly pieced together an account of its formation and the timescales involved, and the tools they use to do this.

In our quest, we journey back to the first moments of the universe, when much of its current composition was established, and we examine the solar system's earliest beginnings in the extremely tenuous matter that lies between the stars. We look at the solar nebula—the cloud of gas and dust surrounding the young Sun that formed a nursery for the growing planets. We describe the origin of each of the planets and other members of the solar system, and how they came to be so different from one another.

The pace of scientific progress has increased over time, and the past two decades in particular have seen a flurry of discoveries and breakthroughs. Inevitably, some gaps in our knowledge remain, and some scientific theories that hold sway today may fall by the wayside tomorrow. Science has a direction, however. New discoveries build upon the work of earlier generations, typically adding to this work rather than demolishing it. Even when a major revolution takes place, its pillars almost always rest on earlier foundations. The rapid recent pace of discovery makes now a good time to take stock of the current situation. While some of the details may change in the coming years, there is every reason to believe that many of the key concepts we describe in this book will survive the passage of time.

Our intention has been to write for general readers who have some basic understanding of science but not necessarily any specialist knowledge of the solar system and its origin. We have tried to avoid using jargon and technical terms where possible, collecting words and concepts that might be unfamiliar in a glossary at the end of the book. We have also taken to heart the maxim that every equation included in a book of this kind is likely to deter far more readers than it will entice.

In researching and writing this book, we have had a good deal of help and support from others. In particular, we would like to thank Conel Alexander, Erik Asphaug, Lindsey Chambers, Mike Edmunds, David Jewitt, Stella Kafka, Lee Macdonald, Simon Mitton, Derek Ward-Thompson, and Iwan Williams for their invaluable contributions. We would also like to thank Ingrid Gnerlich of Princeton University Press for her support, encouragement, and patience.

FROM DUST TO LIFE

ONE

COSMIC ARCHAEOLOGY

A FASCINATION WITH THE PAST

The temple at Karnak on the River Nile is one of the most magnificent monuments to survive from ancient Egypt. Construction of the vast temple complex began 3,000 years ago, and 30 different pharaohs developed and extended the site for a millennium afterward. Everywhere at Karnak, the stone walls and columns of the temple precincts are inscribed with historical texts, prayers, and accounts of religious rituals. Today, guides routinely explain to tourists the meaning of the symbols incised in stone and the significance of this immense monument. Yet for 1,500 years no one in the world could make sense of the writing, and much of ancient Egyptian civilization was a mystery.

The inscriptions at Karnak are composed of hieroglyphics, one of the oldest written languages in the world. The ancient Egyptians used this pictorial script for formal and sacred documents, but its use declined after Egypt became a Roman province in 30 BC. When Egypt became Christian in the 4th century AD, all memory of hieroglyphics was lost. Over the following centuries, scholars puzzled over the meaning of hieroglyphs but never managed to decode them.

In 1799, a French soldier in Napoleon's army discovered a gray slab of stone built into a fort near the Egyptian town known as Rashid or Rosetta. The stone was inscribed with religious proclamations written in three languages: ancient Greek, hieroglyphics, and a more modern Egyptian script called Demotic. Scholars quickly translated the Greek and Demotic writing and realized the same proclamation was repeated in all three languages. Unfortunately, the top portion of the slab had broken away, leaving only 14 lines of hieroglyphs, but these proved to be enough. A painstaking comparison of the languages and some inspired detective work allowed researchers to decode the hieroglyphics for the first time in more than a millennium. The Rosetta stone became the key to unlocking a priceless treasury of information about ancient Egypt and its people.

The story of the Rosetta stone is a good example of how archaeologists can piece together human history by carefully studying rare artifacts that have survived the rigors of time. Occasionally, evidence of the past is staring us in the face just waiting to be identified, like the stone slabs in Karnak. More often the past is buried under debris accumulated over many centuries, as in the legendary city of Troy in Turkey. The past can even be found hiding in the most unlikely of places, such as the details of human history recorded in our genetic code.

Teasing out this information from a variety of sources and grasping its significance is far from easy. It has taken several centuries to develop the tools and know-how that enable today's scientists to interpret clues from the past and turn them into an account of human history. Breakthroughs in archaeology and other sciences often have to wait for a chance discovery like the Rosetta stone, or the introduction of a new technology, or the unique insights of an imaginative mind. Despite these difficulties, scientists persevere because of a deep fascination within all of us: a desire to know about our origins.

Scientists pondering the history of the solar system are much like archaeologists sifting through the sands of Egypt. They bring different methods and tools to the job, but both strive to glean as much as possible from precious relics from the past, and combine this with information deduced from our current surroundings. The distances and timescales may be different but the big questions are the same. Where do we come from? How did we get here? What was the world like in the past? Deciphering the history of the solar system is archaeology on a grand scale. For human society to arise, our species needed to evolve from those that went before. Prior to this, life had to appear on a suitably habitable planet orbiting a long-lived star. Before any of this could happen, our solar system had to take form from the near nothingness of interstellar space. The story of this transformation and how scientists have pieced it together is the subject of this book.

A SOLAR SYSTEM TO EXPLAIN

We start by taking stock of the solar system we see today. The solar system is dominated by a star, the Sun, which contains more than 99.8 percent of the system's mass. Compared to any of the planets the Sun is huge: roughly 1.4 million km (840,000 miles) across, or 109 times the diameter of Earth. The Sun is a rather ordinary star, but "average" is not quite the right word: it is actually brighter and more massive than 90 percent of the stars in our galaxy. The Sun is roughly in the middle of its 10-billion-year life span, neither young nor old, and it has few noteworthy features. It lacks the variability, unusual composition, or excessive magnetic field of some of its more exotic stellar counterparts. From the point of view of life on Earth, this is a good thing: a stable and predictable star provides a pleasant environment for life to flourish.

The average density of the Sun is similar to that of water, but it is largely composed of lighter materials—hydrogen and helium—that are tightly compressed by the Sun's gravity. These two chemical elements make up 98 percent of the Sun's bulk, while all the others contribute the remaining 2 percent, a composition that turns out to be a fair reflection of stars in general. Like other stars, the Sun is made of plasma, an electrically charged gas that reaches temperatures of millions of degrees in the solar interior. Nuclear reactions in the Sun's core provide a continuous source of energy that keeps the Sun shining, and this sunlight provides an important source of heat for Earth and the other planets.

The overwhelming mass of the Sun means that its gravity dominates the motion of all the other members of the solar system. To a good approximation, the Sun lies at the center of the system while every other object revolves around it. Somewhat surprisingly, the Sun accounts for only about 2 percent of the solar system's angular momentum, or rotational inertia. The Sun spins rather slowly, with each rotation taking roughly a month, although the Sun's fluid nature means that different layers in its interior rotate at somewhat different speeds. Most of the rotational energy of the solar system is carried by the planets as they travel around the Sun. This fact has puzzled scientists for a long time and has strongly influenced theories for the origin of the solar system, as we will see in Chapter 3.

The Sun has eight major planets. These follow elliptical orbits around the Sun, all traveling in the same direction—anticlockwise when viewed from above the Sun's north pole. The orbits are almost—but not quite—in the same plane, like concentric hoops lying on a table. With the exception of Mercury and Mars, the orbits are very nearly circular. Mercury and Mars follow more elongated paths—in mathematical terms their orbits are eccentric. The eccentricity of Mars's orbit was an important clue that helped early astronomers understand the motion of all the planets, as we will describe in Chapter 2.

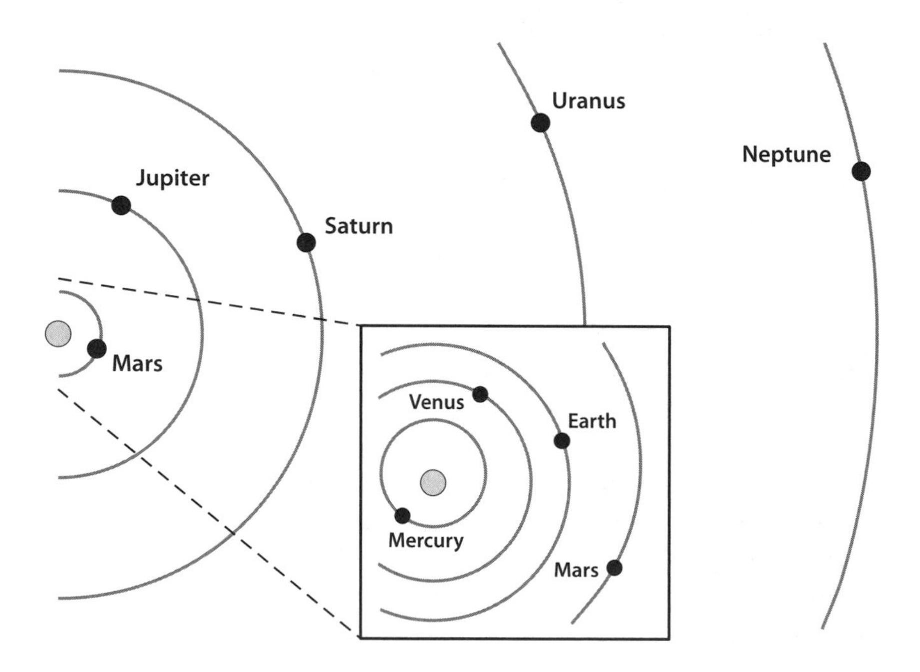

Figure 1.1. The layout of the solar system. The orbits of the major planets are shown approximately to scale.

A useful yardstick for measuring distances in the solar system is the astronomical unit, or AU for short. This is the average distance between Earth and the Sun, roughly 150 million km (93 million miles). The realm of the major planets extends out to 30 AU from the Sun, but it is divided into two distinct domains. The four inner planets all orbit within 2 AU of the Sun. These small objects are called the terrestrial (Earth-like) planets since they all have solid surfaces, and their structure and composition resemble those of Earth.

The four outer planets are arranged more spaciously, orbiting between 5 and 30 AU from the Sun. These bodies are giants compared to the terrestrial planets. Jupiter, the largest, is 300 times more massive than Earth. The giant planets are constructed in a very different way than their rocky cousins, consisting of multiple layers of gas and liquid with no solid surface.

Each of the giant planets forms the hub of a system of rings and a considerable family of satellites. Saturn's spectacular rings are made up of countless chunks of almost pure water ice, ranging in size from a few meters (several feet) down to tiny specks of dust. The rings of Jupiter, Uranus, and Neptune are much darker and less extensive by comparison. As we write, astronomers have found 168 moons orbiting the four giant planets, but it seems almost certain that more will be discovered in the future. In marked contrast, the inner planets have only three satellites—our own Moon and Mars's two tiny companions, Phobos and Deimos. None of the terrestrial planets has rings.

Before we move on to asteroids, comets, and the other members of the solar system, we need to take a moment to describe how astronomers classify things. Astronomical bodies can be grouped in many different ways: based on their shape (roughly spherical or irregular), their composition (rocky or icy), their appearance through a telescope (fuzzy like a comet or a single point of light), or the nature of their orbits. When it comes to planets, however, the popular feeling is that size is the most important factor: a planet is something that is smaller than a star but larger than everything else. The question is how large. Billions of objects orbit the Sun, ranging in size from Jupiter, with a diameter 11 times larger than Earth, down to microscopic grains of dust. Nature has no regard for our habit of allocating objects to particular pigeonholes.

Table 1.1. Selected properties of the major planets

Planet	*Average distance from Sun (AU)*	*Minimum distance from Sun (AU)*	*Maximum distance from Sun (AU)*	*Orbital inclination (degrees)*	*Mass (Earth = 1)*	*Radius (Earth = 1)*
Mercury	0.39	0.31	0.47	7.0	0.06	0.38
Venus	0.72	0.71	0.73	3.4	0.82	0.95
Earth	1.00	0.98	1.02	0.0	1.00	1.00
Mars	1.52	1.38	1.66	1.9	0.11	0.53
Jupiter	5.20	4.95	5.45	1.3	318	11.2
Saturn	9.58	9.04	10.12	2.5	95	9.4
Uranus	19.23	18.38	20.08	0.8	15	4.0
Neptune	30.10	29.77	30.43	1.8	17	3.9

AU = astronomical units.

To a large extent, the dividing line between a major planet and a smaller body is arbitrary, much like the distinction between a river and a stream.

According to the current convention, our solar system has eight major planets. Pluto used to belong to this club, but astronomers recently moved it to a different category based on its similarity to other objects in the outer solar system. This rearrangement didn't please everybody, and Pluto's status remains a topic of debate. With remarkable foresight, astronomer Charles Kowal reflected on the problem of how to define a planet in his 1988 book on asteroids. The largest known asteroid, Ceres, is 952 km (592 miles) in diameter, while Pluto—which was treated as a major planet at the time—is just over 2,300 km (1,400 miles) across. "What will happen if an object is found with a diameter of 1500 km?" Kowal asked. "Will it be called an asteroid or a planet? You can be sure that astronomers will not answer this question until they are forced to!" On this last point he was entirely correct.

The day of reckoning came in 2003 when astronomers discovered four large objects orbiting beyond Neptune. Three of these, Makemake, Haumea, and Sedna, appear to be about 1,500 km (900 miles) in diameter. The fourth, Eris, is roughly the same size as Pluto but about 27 percent more massive. If Pluto is called a planet, then surely Eris should be as well. Should we classify the other three new objects as planets too? What will happen when more large objects are discovered? Will there soon be 20 planets, or 50, or 1,000? It was time for a reappraisal. In a controversial decision, the International Astronomical Union (IAU) voted to create a new class called "dwarf planets," with Pluto, Eris, and asteroid Ceres as founder members. Pluto, formerly a major planet, was redesignated *minor planet* number 124340, reducing the number of major planets to eight.

As of 2012, only five objects have been added to the list of dwarf planets. That still leaves many thousands of known objects that are not planets, dwarf planets, or moons. According to the IAU, these are "small solar system bodies," a category that is divided into "comets," icy bodies that sometimes develop a fuzzy coma and a tail, and "minor planets," rocky objects that always look like points of light when seen from Earth. Few people actually use the term "minor planet" in practice, and small rocky objects are almost always called "asteroids" instead.

A major belt of asteroids lies between the terrestrial and giant planets. Astronomers have found over 300,000 asteroids so far, mostly concentrated between 2.1 and 3.3 AU from the Sun. Hundreds more are discovered every month. Close-up pictures show that asteroids look very different from planets: they are often elongated or have irregular shapes, and their surfaces are covered in ridges, boulders, and craters. Despite their great number, the asteroids contain relatively little mass in total. If all the known asteroids were combined into a single object, it would be smaller than Earth's Moon.

The vast majority of asteroids lie in this main belt between Mars and Jupiter, but some venture farther afield. Asteroid Eros crosses the orbit of Mars, and in 1931 it came within 23 million km (14 million miles) of Earth—about half the minimum distance to Venus. Another asteroid, Hidalgo, moves on a highly elliptical orbit that takes it out beyond Saturn. Some asteroids even cross Earth's orbit, and a small fraction of these will eventually collide with our planet. Two large groups of asteroids, called Trojans, share an orbit with Jupiter, traveling in lockstep around the Sun 60 degrees ahead of the planet or 60 degrees behind it. Astronomers have recently found similar Trojan asteroids that share orbits with Mars and Neptune.

Another belt of small bodies orbits the Sun just beyond Neptune. This region, called the Kuiper belt, is home to Pluto, Eris, and hundreds of other objects found within the past two decades. These discoveries are probably just the tip of the iceberg, and the Kuiper belt probably contains far more mass than the main asteroid belt. Astronomers usually refer to bodies orbiting beyond Neptune as Kuiper belt objects or trans-Neptunian objects to distinguish them from "asteroids," a term that has come to mean small bodies in the inner part of the solar system.

Only a handful of comets have been viewed at close range. These typically look rather like asteroids, although they contain large amounts of ice as well as rocky dust. Comets remain inert as long as they stay cold. However, if a comet comes within a few AU of the Sun, its ices begin to vaporize, releasing gas that blows dust grains off the surface. This gas and dust accumulates around the solid nucleus, forming a huge diffuse cloud called a coma, and streaming away into space to form tenuous tails (one of gas, one of dust) that can extend for millions of kilometers (millions of miles).

Asteroids orbit within a few AU of the Sun, and astronomers had long assumed they were free of ice. In 1996, asteroid Elst-Pizarro surprised many people by developing a tail like a comet as it passed the point in its orbit closest to the Sun (Figure 1.2). In 2001 and 2007, the same thing happened again. Elst-Pizarro is now classed as both a comet and an asteroid. Several other objects in the outer parts of the asteroid belt display this dual personality. These bodies must harbor reservoirs of ice that partially vaporize when the temperature becomes high enough. Icy deposits have recently been detected on the surface of Themis, one of the largest asteroids in the main belt. It may be that other asteroids contain ice in their interior, protected from sunlight by a layer of rocky dust on the surface. Clearly, the boundary between asteroids and comets is not as sharp as astronomers once believed.

Most comets follow highly elongated orbits, arriving in the inner solar system from beyond Neptune and then making the return journey. A few hundred comets have become trapped on smaller orbits by the pull of Jupiter's gravity, and these rarely travel much beyond the giant planet's orbit. Typically, these "Jupiter family comets" have traveled around the Sun many times, losing much of their former glory over time. Most comets move on much larger orbits by comparison, taking thousands or even millions of years to travel around the Sun. Tracing the motion of these "long-period" comets backward in time along their orbits shows that they come from a vast reservoir of icy bodies far from

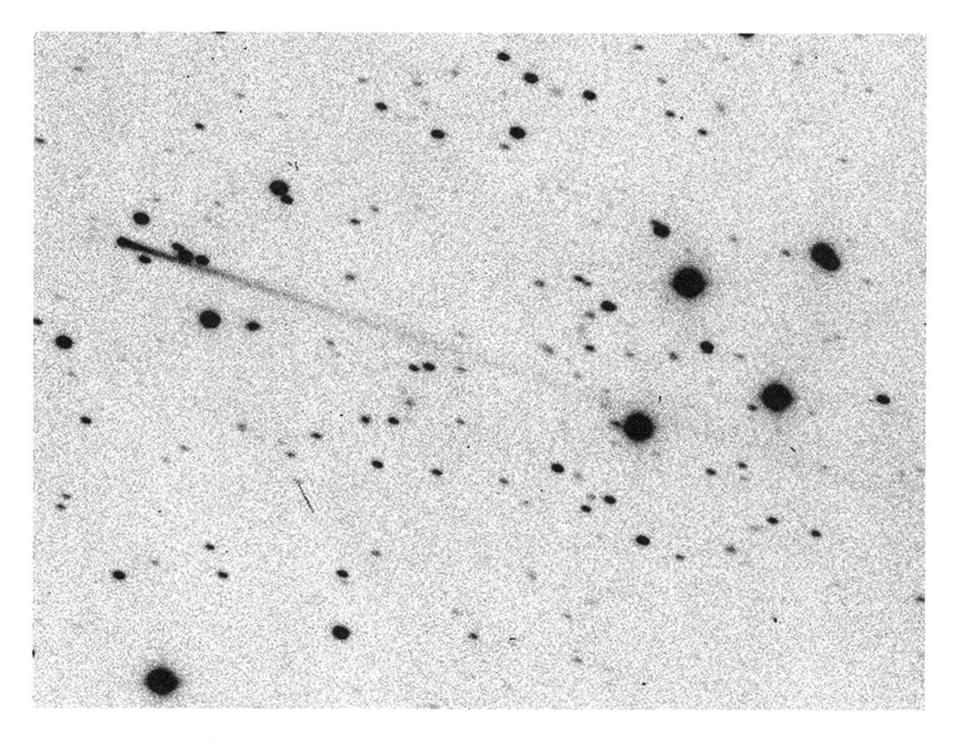

Figure 1.2. Comet Elst-Pizzaro. From this image taken by Guido Pizarro at the European Southern Observatory on August 7, 1996, and follow-up images taken later the same month, Eric W. Elst was able to identify the comet displaying a narrow dust tail with the asteroid 1979 OW7 discovered on July 24, 1979, when it had no tail. (ESO)

the Sun. This spherical swarm of comets, known as the Oort cloud, is concentrated between 20,000 and 50,000 AU from the Sun, and it marks the true outer boundary of the solar system.

REAL WORLDS

Any successful scenario for the origin and evolution of our solar system needs to account for the overall structure of the planetary system. It also needs to explain the nature of individual objects, including features that are readily apparent such as the cratered surface of the Moon, and information buried deep within planetary interiors. For centuries, astronomers had little on which to base their theories. Most objects in the solar system appeared as tiny circles or points of light through a telescope. Even today, the best telescopes cannot obtain images or data as detailed as those from a passing spacecraft.

The dawn of the space age marked a dramatic turning point in how we view the solar system. Space flight allowed astronauts to visit the Moon and bring back 382 kg (842 pounds) of lunar rocks, prompting a burst of new research on Earth's nearest neighbor. Space missions also transformed many hazy images and tiny points of light into real worlds that could be mapped, probed, and studied scientifically, providing vast amounts of new data.

The Mariner 4 mission to Mars demonstrates that when it comes to exploration there is no substitute for going there. The Mariner 4 spacecraft was launched from Cape Canaveral, Florida, in November 1964, carrying a television camera and half a dozen science instruments. If all went well, Mariner 4 would take the first close-up pictures and measurements of any body in the solar system apart from Earth and the Moon. Unfortunately, six previous attempts by the United States and the Soviet Union to reach the red planet had all failed, including Mariner 4's sister ship, Mariner 3. No one knew if Mariner 4 would meet the same fate.

As Mariner 4 approached Mars in July 1965, its progress was followed eagerly by scientists and the public alike. Mars was the only rocky planet whose surface had been seen by astronomers on Earth, but the view through a telescope was frustratingly fuzzy. Astronomers knew that Mars has white polar caps like Earth, and shifting areas of light and dark terrain, but they could see little else from afar. Many people anticipated that Mars would be a cooler, miniature version of our own planet. Percival Lowell's wilder speculations at the end of the 19th century, envisaging a Martian civilization that had built a network of canals across the surface of the planet, had long since been discredited. However, the existence of life, particularly vegetation, was regarded as a serious possibility. The prospect of finding life elsewhere in the solar system was a key driver behind NASA's fledgling planetary exploration program and an important source of public and political backing.

Mariner 4 flew past Mars on July 14 and 15, 1965, at a distance of only 9,846 km (6,118 miles) above the surface. The spacecraft captured 22 rather hazy, black-and-white TV images, which it stored on a tape recorder and later transmitted back to Earth. These first indistinct snapshots of Mars covered only 1 percent of the planet's surface, but their effect was stunning. One contemporary journalist proclaimed that there had been "no comparable discovery since Galileo turned his telescope on the Moon."

The fantasy image of an Earth-like Mars, built up over the previous century, was shattered in a day when the TV images arrived. The real Mars displayed a heavily cratered terrain that was much more reminiscent of a desolate lunar landscape than our home planet (Figure 1.3). This picture of a hostile and alien environment was reinforced by measurements that put the surface temperature at –100°C (–150°F) and the atmospheric pressure at less than one-one hundredth that on Earth. It later turned out that the cratered terrain seen by Mariner 4 is not representative of Mars as a whole, but the preconceptions formed in the era before space flight were overturned forever. Hopes of finding life in the solar system had been dashed for the time being, but there were still good reasons to explore further. If Mars could spring such a surprise, what might we find on other planets and moons?

As we write in 2012, space missions have flown past every major planet in the solar system. Orbiting spacecraft have mapped and studied Venus, Mars, and the Moon in great detail, as well as Jupiter and Saturn and their systems of rings and moons. The Messenger spacecraft entered orbit around Mercury in 2011 to perform a similar survey of the planet closest to the Sun. Robotic probes have landed on Venus, Mars, and Saturn's moon Titan, returning images and data from the surface, and astronauts have collected samples from our Moon. Spacecraft have traveled to several asteroids and comets, and the New Horizons spacecraft, on its way to the Kuiper belt, is due to pass close to Pluto in 2015. These space missions, together with observations using the Hubble Space Telescope and a new generation of ground-based telescopes, have allowed scientists to compare different planetary worlds in detail for the first time and to address fundamental questions about how they formed.

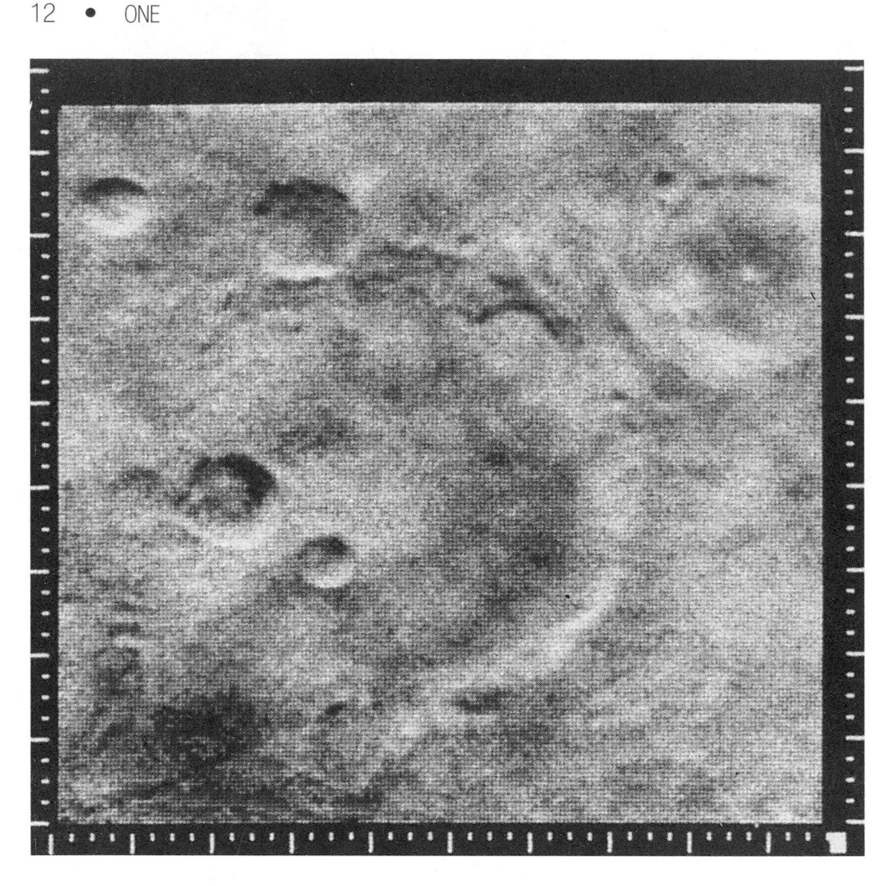

Figure 1.3. One of the images of Mars returned by the Mariner 4 spacecraft in July 1965. Filling most of the frame is a somewhat indistinct crater 151 km (94 miles) in diameter, which was named Mariner after the spacecraft. (NASA/JPL)

WINDING BACK THE CLOCK

We now know that every object in the solar system has a unique identity that reflects its formation and evolution over the age of the solar system. Cosmic history is deeply etched into these worlds: in their composition, in their structure, and in the orbits they follow. Interpreting these clues to compose a history of the solar system requires a good deal of detective work involving many scientific disciplines including physics, chemistry, geology, and astronomy. Several scientific principles and practical techniques central to such work crop up repeatedly, so we will describe them briefly now before beginning the main story.

One recurring theme is the effect of heating and cooling. We all know that living organisms are highly sensitive to temperature, but planetary materials also respond markedly when heated or cooled, often in ways that are permanent. To give an example, imagine a rocky planet that is heated by asteroids colliding with its surface or by the decay of radioactive materials in its interior. As the planet grows hot, its rocks will begin to melt. If enough of the planet melts, denser materials such as iron will sink toward the planet's center while lighter materials float upward. Later, when the heat source dies away, the planet will cool and solidify, and new rocks will form. The kind of minerals that form within these rocks depends on the temperature and pressure, as well as how fast the rocks cool, and whether or not the planet separated into layers. All this information is imprinted in a planet's rocks and can survive for billions of years to be interpreted by modern experts.

Spacecraft have landed on only a handful of bodies in the solar system, and objects beyond the solar system lie far out of reach. Luckily, nature allows us to find out what an object is made of simply by observing it from a distance, from either a passing spacecraft or with a telescope on Earth. The light from a star, a planet, or any body either emitting or reflecting light can be split into its component colors to form a spectrum. A typical stellar spectrum contains thousands of dark, narrow gaps, called "lines," where atoms of the various chemical elements in the star's outer atmosphere have absorbed light, each at a characteristic set of wavelengths. The amount of absorption is related to the abundance of the element responsible, so it is possible to work out the star's composition using these lines. The spectra of planets and asteroids are somewhat harder to interpret since such bodies contain molecules and minerals that form broader absorption features than atoms in stars. Still, it is often possible to deduce a good deal about their composition from their spectrum. The same kind of analysis can be applied in "invisible" regions of the spectrum, such as infrared light.

Radioactivity also plays a central role in our story. Naturally occurring radioactive elements are incredibly useful tools for examining the past because they have built-in timers. When a radioactive substance is incorporated into a mineral or a living organism, or even the solar system as a whole, it behaves as if a stopwatch has been activated. The amount of radioactivity decreases in a predictable manner, falling by half in a fixed period of time, called the half-life, which is unique to each radioactive material. After two half-lives, only a quarter of the radioactive material remains, one-eighth is left after three half-lives, and so on.

When a radioactive element decays, it typically changes into another element, often one with very different physical and chemical properties, allowing the decay process to be clearly identified. By measuring how much radioactive material is left, and its distribution within an object, scientists can tell when the object formed. (We will see how this is achieved in Chapter 4.) Even after all the radioactivity has disappeared, the distribution of the decay products often tells us something about the early history of the object. This technique, called radiometric dating, works on any timescale from centuries to billions of years as long as a radioactive material with a suitable half-life can be identified in the sample. Radiometric dating is equally useful to archaeologists studying a wooden coffin from ancient Egypt and chroniclers of the solar system measuring the age of rocks from the Moon.

Another tool scientists use to reconstruct the past is numerical modeling. We would like to be able to wind back the clock and watch the solar system while it was forming and evolving to its current state. Of course, that is impossible in reality, but an approximate way to do it is to use a computer model—a kind of virtual reality that simulates the solar system or some of its members. A model consists of a set of mathematical equations that encapsulate the known laws of physics and properties of materials measured in the laboratory, together with a snapshot of the system at some point in time.

A simple model might begin with Newton's law of gravity, add the positions, speeds, and direction of motion of the planets, and ask how the planets will move over the next 100 years. More complicated models could include collisions between objects and calculate their thermal and chemical evolution over time. This kind of modeling has helped to revolutionize our thinking about the formation and evolution of the solar system, allowing scientists to test and refine complicated theories in ways that could not be done otherwise. Models are particularly useful in situations that can't be studied in a laboratory, such as a collision between two planet-sized bodies, or to examine the behavior of materials over millions of years. However, computer models are only as good as the data we put into them. Models can help make sense of the information we gain by observation and experiment, but they can never replace these things. We are still a long way from being able to program a computer to tell us exactly what happened in the past.

Astronomers can also wind back the clock by looking at other stars and planetary systems that are younger than our own. These are not exact replicas of the solar system, but we can still learn a good deal about how planetary systems form and evolve by looking at younger systems in various stages of development.

Many newborn stars are surrounded by disk-shaped clouds of gas and dust that seem to be evolving into planets. By carefully measuring the size, structure, and composition of these disks, astronomers are putting together a picture of what our own solar system looked like while it was forming. For the first time in human history, we also have examples of other fully formed planetary systems to study. In 1992, Alexander Wolszczan and Dale Frail discovered two planets orbiting a pulsar—a rapidly spinning dead star. Three years later, Michel Mayor and Didier Queloz announced the first indisputable detection of a planet orbiting an ordinary Sun-like star (51 Pegasi). By late 2012, the number of stars known to have planets exceeded 500. Systems of two or more planets had been identified orbiting more than 60 of these stars. The total planet count was over 800 and rising rapidly. The Kepler space mission has identified more than 2,000 stars that appeared to have planets, and work continues to confirm or reject the planet "candidates."

Many of these extrasolar planets are gas giants like Jupiter, or they lie very close to their stars, making them too hot to support life. The explanation for this is that the discovery process has been biased by the techniques astronomers have had available to find them—large planets and planets that orbit close to their star are simply the easiest to find. However, the picture is changing rapidly as technology improves, and astronomers are starting to find planets that are similar to Earth in size and may resemble our own planet in other ways as well.

The discovery of extrasolar planets means that scientists are no longer limited to studying a single planetary system—our own. Instead we have literally hundreds to choose from. The properties of other planetary systems are helping to shed light on how our own solar system formed and evolved. For example, it seems unlikely that planets orbiting very close to their star could have formed where they are today. The discovery of such objects led to the realization that planets can migrate far from their birthplace. As we will see in Chapters 9 and 14, researchers are now examining whether the planets in our own solar system could have migrated substantial distances after they formed.

PUTTING THE PIECES TOGETHER

Logically, as we learn more about our solar system and other planetary systems, it should get easier to work out how these systems formed. In one sense this is true. When we know little, there is often no way to challenge simple, plausible-sounding ideas that are false. Long ago, for example, it seemed entirely reasonable that Earth is static while the Sun, the planets, and the stars all revolve around us. As more and better data accumulate, theories that don't fit observations have to be discarded. New information forced our ancestors to accept that Earth is not the center of the universe, and that our planet is spinning and hurtling through space. However, a wealth of data can also make life more complicated for scientists because any successful theory has to explain many more things.

In 1796, the French mathematician Pierre Simon Laplace devised one of the earliest scientific scenarios for the origin of the solar system, which we will explore in greater detail in Chapter 3. Laplace based his work on the handful of facts available to him at the time, and his "nebular hypothesis" was simplistic and short on details as a result. What he achieved was rather like putting together the dozen chunky pieces that make up a toddler's jigsaw puzzle. Today, scientists have collected vast amounts of information from every corner of the solar system. Making sense of all these data has become equivalent to assembling the most difficult jigsaw imaginable, with thousands of tiny pieces to be slotted into place. What's more, we don't know whether a few vital pieces are still missing. Perhaps we still need a Rosetta stone that will explain the meaning of numerous other observations and allow us to clearly see the history of the solar system for the first time.

The European Space Agency must have had this thought in mind in 1993 when it gave the go-ahead for the first space mission to orbit around a comet and land a probe on its surface. At the time, it was widely believed that comets are like time capsules—repositories of pristine material that have survived unchanged and uncontaminated since the dawn of the solar system. The astronomers planning the mission hoped that studying a comet at close quarters would unlock many of the secrets of the early solar system. To reinforce this ambition, they boldly named the mission Rosetta after the famous stone from Egypt.

The Rosetta spacecraft finally embarked on a 10-year journey to comet Churyumov-Gerasimenko in March 2004. By the time the spacecraft reaches its goal in 2014, the concept for the Rosetta mission will be more than 20 years old. Over the intervening years, scientists have begun to question whether comets are quite so pristine as they had once imagined. In 2004, for example, NASA's Stardust spacecraft flew past comet Wild 2, scooping up a precious sample of dust that it later returned to Earth. When scientists examined this dust, they discovered that some of it had once been heated to temperatures as high as 1400°C (2500°F), quite unlike the frozen, primeval matter comets were supposed to contain. In 2010, the Deep Impact spacecraft visited comet Hartley 2, producing several surprises. Unlike most comets, Hartley 2's activity is driven by the evaporation of carbon dioxide rather than water ice. Even more strange, the comet's nucleus consists of two lobes with different chemical compositions (Figure 1.4). The two parts appear to have formed at different distances from the Sun and later coalesced to become a single comet.

It seems that comets, like planets, moons, and asteroids, are individuals each with its own unique and complex history. If Rosetta completes its daring mission, we will surely learn much as a result. Unfortunately, the planners' desire to unravel the mysteries of the solar system by studying a single comet is unlikely to be realized for now.

Scientific inquiry by its nature is continually in a state of flux as new information becomes available and theories are refined. However, the process has a direction. Scientists, in their role as cosmic archaeologists, are moving ever closer to understanding the solar system's past and how it came to be. The pace of discovery has been especially rapid over the past two decades. The jigsaw is not yet complete, and a few of the pieces may be in the wrong place, but we can now see enough of the picture to make the story worth telling. In the next chapter, we begin this story by examining how astronomers came to appreciate the solar system in all its diversity.

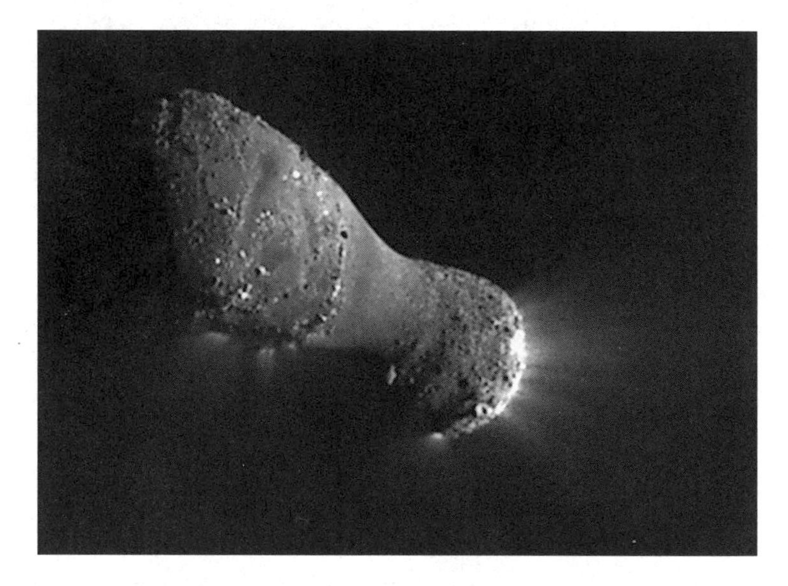

Figure 1.4. Jets streaming from the nucleus of Comet Hartley 2, which is about 2 km (1.2 miles) long. The image was taken in November 2010 from a distance of about 700 km (435 miles) by NASA's Deep Impact spacecraft while on an extension to its initial mission. (NASA/JPL-Caltech/UMD)

TWO

DISCOVERING THE SOLAR SYSTEM

MEASURING THE SOLAR SYSTEM

In 1768, James Cook and a party of 80 men set sail from England on a voyage of astronomical discovery that would take them halfway around the world. After 10 months at sea, HMS *Endeavour* and its crew arrived at their destination, the remote island of Tahiti in the middle of the Pacific Ocean. Their main mission was to spend a mere six hours making astronomical observations that could help to establish the true size of the solar system once and for all.

Cook was charged with observing a rare phenomenon called a transit of Venus—the passage of Venus's silhouette as it moves directly in front of the Sun (Figure 2.1). Astronomers had laid elaborate plans for the transit on June 3, 1769, to be watched simultaneously from several locations around the world. All the observers would time how long the transit lasted. By combining these measurements, the astronomers hoped to calculate the precise distance between Earth and the Sun—one astronomical unit—and use it to establish the scale of the solar system. Transits of Venus are incredibly rare because the Sun, Venus, and Earth all have to line up perfectly. The next one would not take place for more than a hundred years. Everything depended on good weather—a sky covered in clouds would mean that Cook's long journey had been wasted. As luck would have it, the weather in Tahiti on June 3 was perfect. Cook wrote in his diary, "The day proved as favourable to our purpose as we could wish; not a cloud was to be seen the whole day, and the air was perfectly clear."

Figure 2.1. The transit of Venus of June 5–6, 2012, as seen from Earth orbit in ultraviolet light by the Solar Dynamics Observatory. (NASA/SDO)

By the time of Cook's voyage, astronomers knew how to calculate the relative spacing of the Sun, the planets, and their satellites by tracking their movement across the sky, but the actual distances were not known with any certainty. Astronomers found this highly unsatisfactory. How could they begin to understand the solar system if they were not even sure how big it was? Securing an accurate value for the astronomical unit (AU) was a high priority. Scientists considered this issue to be so important that they launched the most ambitious international collaboration of its day to settle the matter. Cook set out on the first of his three epic voyages of discovery as part of the venture.

The first scientifically rigorous attempt to measure the size of the solar system had been made a century earlier in 1672. Jean-Dominique Cassini, director of the Paris Observatory, dispatched his fellow astronomer Jean Richer to observe the position of Mars from French Guiana in South America, while Cassini remained in Paris to make an identical measurement. Seen from these two viewpoints, Mars appeared to lie in a slightly different position in the sky relative to the more distant stars.

Armed with the two positions of Mars and the distance between the observers, Cassini and Richer could work out how far Mars was from Earth. Once they knew this distance, they could calculate every other distance in the solar system—but only in terms of Earth's radius. Two years earlier, Jean-Felix Picard had made the first accurate measurement of Earth's radius, roughly 6,300 km (3,900 miles). Using this number, Cassini and Richer estimated that the Sun lay at a distance of 21,700 Earth radii, or 138 million km (86 million miles).

Then, in 1679, Edmond Halley (of comet fame) realized that transits of Venus provided an unrivalled opportunity to measure the Sun's distance accurately. Using an idea voiced a few years earlier by Scottish mathematician James Gregory, Halley developed a scheme that involved the seemingly simple task of timing the duration of a transit. The catch was that the same measurement had to be carried out simultaneously at widely separated latitudes on Earth. In 1716, Halley published his plan, knowing he would not live to see it fulfilled. Transits of Venus occur in pairs eight years apart, with each pair separated by over a century. At the time, Halley was already 60 years old and the next transits would not take place until 1761 and 1769.

As the date of the transits grew near, astronomers were spurred into action. Cook's voyage was part of a much greater enterprise involving dozens of astronomers who fanned out across the globe to observe the transits in both 1761 and 1769. Members of some expeditions endured years of hardship, misfortune, and even death in the cause of science. When the observations were collected and analyzed, the figure they gave for the astronomical unit was 153 million km (95 million miles), only slightly larger than today's figure of 150 million km (93 million miles).

The solar system turned out to be much larger than previously imagined. In the 2nd century AD, Greek astronomer Claudius Ptolemy, working in the Egyptian city of Alexandria, put the distance of the Sun at a mere 1,210 Earth radii (8 million km or 5 million miles in modern units), an estimate that had stood for 1,500 years. In the 17th and 18th centuries, astronomers were confronted with a solar system almost 20 times larger. These centuries were a time of great scientific upheaval spurred by new instruments and techniques, and the development of powerful mathematical tools. Not only were astronomers coming to appreciate the true scale of the solar system, their entire picture of the solar system and the universe was being turned on its head. Beliefs about the solar system that had gone almost unchallenged for more than a thousand years were being called into question or overturned completely.

FROM WANDERING GODS TO GEOMETRICAL CONSTRUCTIONS

Around the year AD 145, Ptolemy produced a monumental work, popularly known as the *Almagest*, which was a grand compilation of Greek astronomical ideas and tools for computing the positions of the heavenly bodies. Ptolemy's vision of the universe, like that of almost all his predecessors, was geocentric. In Ptolemy's view, Earth lay at the center of the universe, while the Moon, Mercury, Venus, the Sun, Mars, Jupiter, and Saturn all traveled around it on successively larger orbits. These seven bodies were called "planets" (wanderers) because they moved with respect to the fixed constellations of stars.

Even before the emergence of Greek astronomy, Babylonian observers had recorded the positions of the planets on clay tablets as early as 1500 BC. However, the Babylonians were primarily interested in astrology rather than astronomy. To them, the planets were not just wandering lights in the sky, they were gods whose movements foreshadowed events on Earth. The Babylonians recorded long series of accurate observations in the hope of finding repetitive patterns that would enable them to predict the future. Greek natural philosophers adopted the practice of associating the planets with gods. Translated into Latin, these names have come down to us unchanged: Venus, goddess of love, Mars, god of war, and so on.

The astronomers of ancient Greece were the first to describe celestial phenomena in mathematical terms. They imagined a three-dimensional, dynamic universe, but it was a universe based on philosophical preconceptions and the limited observational precision available at the time.

Scientifically speaking, this still was a step forward from the mythology of earlier ages. It was the early Greek thinkers who first used the term "cosmos" for the universe, meaning a unified, harmonious whole, in contrast to chaos.

Of all the Greek philosophers, Aristotle (384–322 BC) had the most profound effect on how the early Western world thought of the solar system. To Aristotle, it was a matter of common sense that Earth should be at the center of the cosmos, while the heavenly bodies traveled around it on circular paths. Earth, and everything else below the Moon, was imperfect, while the heavens beyond were perfect and unchanging.

The difficulty with this point of view was how to explain the motion of the planets in simple mathematical terms. For example, although planets generally move across the sky in the same direction, their speed varies and occasionally they go into reverse, performing slow loops called retrograde motion. The brightness of the planets also varies over time, and the apparent sizes of the Sun and Moon change as well. Why the planets should behave in such an odd manner was a great puzzle to Greek philosophers.

Today, we understand this behavior because we know that the planets travel around the Sun rather than Earth, and that all the orbits, including the orbit of the Moon around Earth, are elliptical rather than circular. It is much more difficult to explain things in terms of circular motion centered on Earth. With considerable mathematical ingenuity, Greek astronomers were able to reproduce retrograde motion and the varying speeds of the planets using complex combinations of circles moving around other circles with their centers displaced from Earth. These methods themselves generated puzzles and paradoxes, but they were tolerated in the absence of another acceptable theory. The leap of imagination needed to accept a Sun-centered system was too great for most people. Aristarchus of Samos, in the 3rd century BC, was one of the few philosophers who taught that Earth orbits around the Sun instead of the other way around. He found few followers.

Aristotle embraced the ideas of his near contemporary, Eudoxus of Cnidus, who imagined that the planets were carried along by transparent celestial spheres, nested inside one another with their axes offset. For

Aristotle, these rotating spheres were as real as the planets themselves. The ultimate driver was an outer sphere carrying the stars to which the spheres of the planets were linked. In this arrangement, there was no reason to suppose that the stars were significantly farther from Earth than the planets themselves. Building on Eudoxus's ideas, Aristotle developed a model requiring 56 spheres in total. Despite this complexity, Aristotle's model still failed to match observations in detail.

By the time Ptolemy wrote his *Almagest*, some 500 years later, he was more interested in finding a way to accurately predict the motions of the planets than in constructing a physical description of the cosmos. Ptolemy readily combined inconsistent mathematical tools to explain different features of the same body, such as the speed and apparent size of the Moon. Clinging doggedly to the principle of circular motion, Ptolemy was forced to combine and displace his circles to fit the observations. In his model, planets did not move in a simple circular orbits around Earth but made additional circular loops, or epicycles, as they went. Ptolemy also displaced each whole looping orbit so that its center circled around Earth at some fixed distance. But this was still not enough. To account for the varying speed of the planets, he had to assume that a planet's speed was uniform not as measured from the center of Earth, or even from the displaced center of the looping orbit, but from another point in empty space called the "equant." In doing this, Ptolemy essentially abandoned the concept of uniform circular motion. However, the idea of an Earth-centered universe was so entrenched that Ptolemy remained the ultimate authority on astronomy for more than a millennium.

The development of Greek astronomy ceased after Ptolemy, but the works of the Greek philosophers and astronomers were discovered by the emerging Islamic world after the Islamic conquests of the Middle East and Mediterranean lands, beginning in the 7th century. They were translated into Arabic in Baghdad in the early 9th century. From there, they spread to North Africa and Spain and were later translated into Latin. The basic tables used to predict the positions of the planets were improved, but the underlying geocentric picture remained (Figure 2.2). In medieval Europe, Aristotle's philosophy was merged with concepts from Christian theology. Backed by the weight of the Catholic Church, the geocentric view of the cosmos became doubly difficult to question.

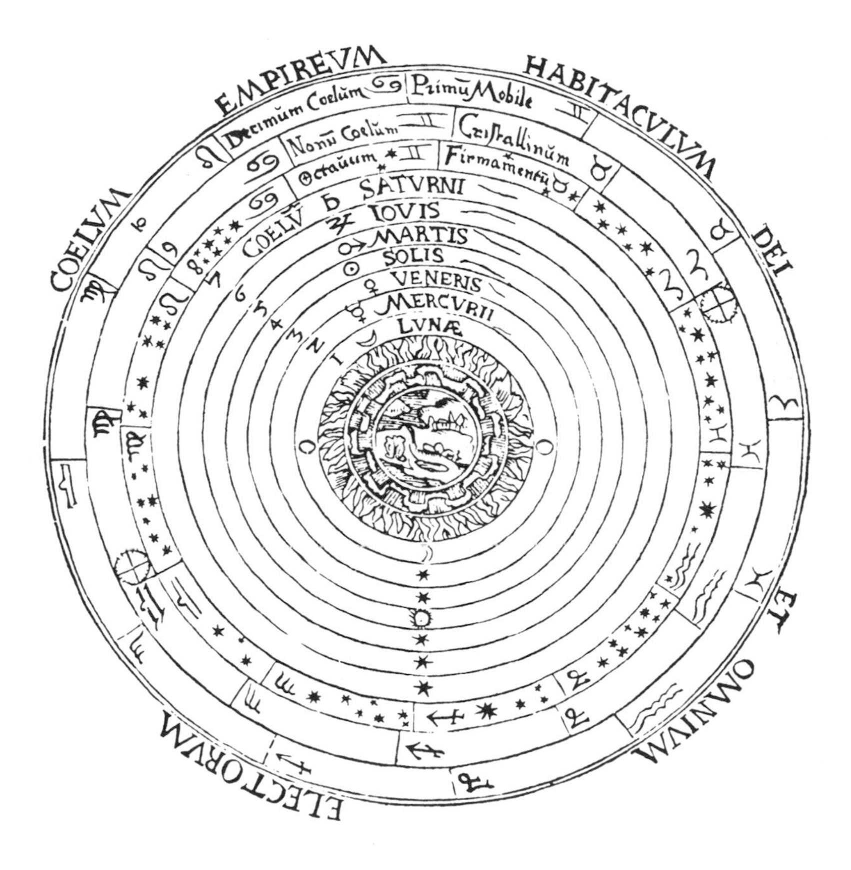

Figure 2.2. A 16th-century representation of the universe with Earth at the center, from a 1539 edition of *Cosmographia* by the German astronomer and cartographer Peter Apian. (J. Mitton)

THE SUN TAKES CENTER STAGE

The first person to voice serious doubts about Ptolemy was Nicolaus Copernicus. Born in the Polish city of Torun in 1473, Copernicus became an accomplished Renaissance scholar, studying at the universities of Krakow, Bologna, and Padua. He qualified in law and medicine and in 1501 settled into a lifelong administrative career at the Cathedral of Frauenberg. However, Copernicus's early studies at Krakow had stimulated an interest in astronomy and, between 1510 and the early 1530s, he developed a heliocentric theory of the universe—one centered on the Sun rather than Earth.

Copernicus was driven to the conclusion that Earth orbits the Sun by a deep dissatisfaction with the inconsistencies and arbitrary assumptions in Ptolemy's mathematical rules for the solar system. These rules made no physical sense to Copernicus and could not describe reality. Copernicus wanted to return to the philosophy of Aristotle and find a true system that described the world and explained the observations. Like Aristotle, he favored uniform circular motions and particularly objected to Ptolemy's use of equants. "Being aware of these defects," he wrote, "I spent much time considering whether one might perhaps find a more reasonable arrangement of circles from which every apparent inequality could be calculated and in which every element would move uniformly about its own center, as the rule of absolute motion requires."

Putting the Sun at the center of the planetary system, having the Moon orbit Earth, and letting Earth rotate once each day seemed like obvious solutions to the problems that beset Ptolemy's model. In Copernicus's opinion, the idea that Earth is moving was not contrary to Christian doctrine, although he was aware that some might take a different view. But when Copernicus came to work out the mathematics and fit the observations, he discovered that the devil was in the detail. Uniform circular motion could not account for the varying speeds of the Moon and planets, and Copernicus found he still needed to use epicycles or eccentric orbits.

Copernicus's book, *On the Revolutions of the Heavenly Spheres*, was finally published in 1543, the year of his death, thanks to the encouragement and intervention of a younger disciple, Georg Joachim Rheticus. As a practical manual for predicting the positions of the planets, it was little advance on Ptolemy. But Copernicus had changed the map of the known universe. For the first time, Copernicus had described a *solar* system—a system centered on the Sun, not Earth. Copernicus put the planets in their correct order of distance from the Sun: Mercury, Venus, Earth, Mars, Jupiter, and Saturn (Figure 2.3). He also realized that, because the constellations didn't seem to change, even as Earth traveled around the Sun, the stars had to be at least several hundred times farther away than Saturn.

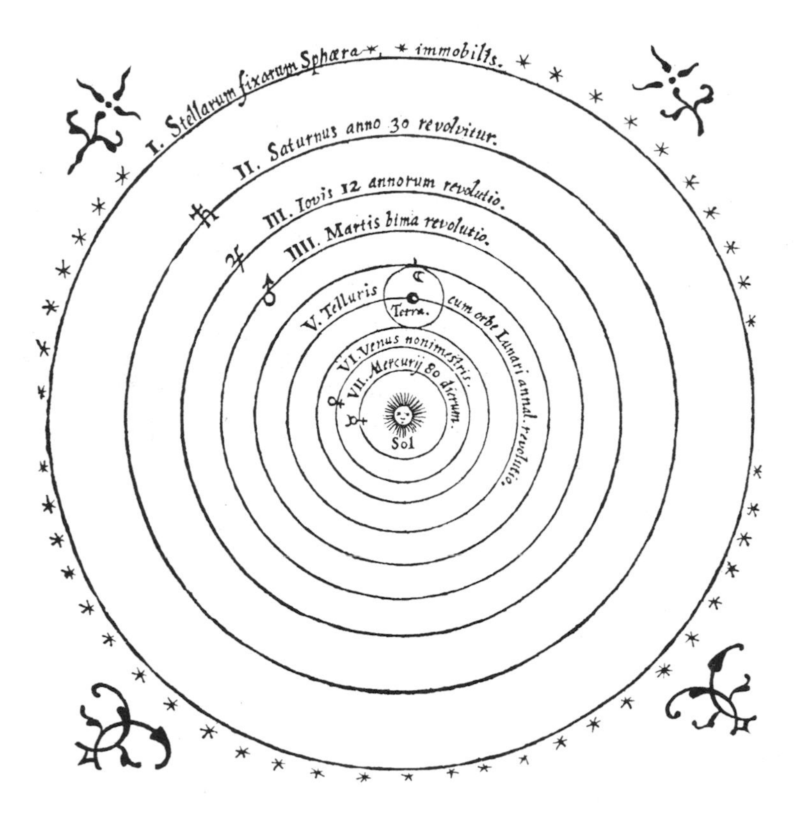

Figure 2.3. The solar system, as pictured by Nicolaus Copernicus in his 1543 book, *On the Revolutions of the Heavenly Spheres.* (J. Mitton)

LAWS AND ORDER

For half a century, Copernicus's visionary work made little impact. One astronomer who was convinced was Johannes Kepler. Born near Stuttgart in 1571, Kepler studied at the University of Tübingen and became a teacher of mathematics at Graz in 1594. Copernicus's heliocentric model appealed to Kepler because he believed the Sun was the source of the harmony he saw in the arrangement of the solar system. In particular, he thought the Sun's central location was responsible for moving the planets along their orbits. This was a radically new way of thinking about the cosmos. In Kepler's view, the planets moved for reasons that were physical rather than philosophical.

In 1596, Kepler's work came to the attention of the influential Danish astronomer Tycho Brahe, who also had doubts about Aristotle's picture of the universe. In 1572, when Tycho was in his 20s, he had witnessed the temporary appearance of a new star in the constellation of Cassiopeia, an outburst that we would call a supernova today. Tycho calculated that this new star lay far beyond the Moon, proving that the supposedly unchanging heavens were capable of change after all. Five years later, Tycho proved that the great comet of 1577 was moving through the realm of the planets and that comets were not an atmospheric phenomenon as previously believed. This too ran counter to Aristotle's notion of a perfect unchanging cosmos.

Sponsored by the king of Denmark, Tycho built an elaborate observatory equipped with high-quality instruments that he designed and made himself. For 20 years, Tycho worked tirelessly making astronomical observations of unprecedented accuracy on a scale never before attempted. In 1597, when his Danish sponsorship waned, Tycho moved to Prague to become the official astronomer to the Holy Roman Emperor, Rudolph II. It was here, three years later, that he met Kepler, 25 years his junior. Supplied with some of Tycho's observations, Kepler began work trying to deduce the shape of Mars's orbit. When Tycho died the following year, Kepler inherited a rich legacy of data as well as Tycho's job as imperial astronomer.

Kepler tried and discarded several theories for Mars's motion before he finally cracked the problem: Mars's orbit around the Sun could be described almost perfectly by an ellipse with the Sun located at one focus. He immediately proposed that all the planets move along elliptical orbits, the first of his three laws of planetary motion. (Laws 2 and 3 describe how a planet's speed varies along its orbit and show how the size of a planet's orbit is linked to its orbital period.) Kepler published his findings in 1609 in a book, *New Astronomy Based upon Causes, or Celestial Physics by Means of Commentaries on the Motions of Star Mars*. From this moment on, circular motion was consigned to history.

In the same year, news from Holland of the invention of the telescope reached Galileo Galilei in Padua, Italy. Galileo began work constructing his own telescopes, and in the winter of 1609–10 he started to observe the night sky as never before. What he discovered dispelled any lingering doubts in his own mind about Copernicus's heliocentric system. Galileo found a lunar landscape covered in mountains and plains, like Earth and most unlike the perfect celestial orb envisioned by Aristotle. Venus exhibited phases like the Moon but in a way that meant it couldn't be traveling around Earth. Jupiter was accompanied by a family of four satellites that traveled around the planet, proving that Earth was not the center of everything.

Galileo's subsequent battle with the Catholic Church is well known. For many scholars of the time, the boundary between theology and scientific inquiry was blurred, and both had to be considered in the search for truth. Galileo succeeded in antagonizing many religious authorities of the day, and his advocacy of an Earth in motion appeared to be at odds with several passages in the Bible. The Church came to regard Galileo as a challenge to its authority and was not ready to endorse such a major change in thinking without stronger proof. After all, no one could explain why Earth's motion is imperceptible if we are indeed hurtling through space. Today, modern scientific instruments can indeed detect this motion, but the effect is subtle, which is why we are usually not aware of it. Galileo never managed to demonstrate Earth's motion directly, but his many other discoveries were sufficiently compelling to spread the notion of a Sun-centered universe throughout Europe despite official censorship.

GRAVITY RULES

In the year in which Galileo died—1642—Isaac Newton was born. Newton's theory of gravitation completed the search begun by Copernicus and continued by Kepler and Galileo, for a plausible physical theory to explain the motion of the Moon and planets. Newton noted that a planet could be made to follow a curved orbit instead of a straight line as a result of a force coming from another body. This force, he realized, was the same "gravitation" that made an apple fall toward the ground here on Earth. If gravity were a universal phenomenon, it would explain the motion of the Moon around Earth and the planets around the Sun, and even the ocean tides. Tackling the mathematics in detail, Newton was able to show that a gravitational force between two objects declining with the square of their separation would explain all of Kepler's laws.

Newton's reasoning required him to make a number of assumptions. One of these was that the space between the planets was essentially empty, a view not held by everyone. The influential French philosopher René Descartes, writing in 1644, had envisaged a universe full of swirling whirlpools of invisible matter, each centered on a star. The Sun lay in one such vortex, he believed, which swept the planets around with it. In Newton's universe, gravity allowed objects to interact with one another across empty space with no physical substance connecting them together.

Newton's theory of gravitation was published in 1687 in *Mathematical Principles of Natural Philosophy*, commonly known by its shortened Latin title, the *Principia*. Although Newton gave a name and character to the force controlling the dynamics of the solar system, he did not attempt to explain how the solar system came into existence nor what started the planets in motion. He was content to believe that creation was the work of God. But one thing troubled him. How could the planets remain on their orbits around the Sun if they were continually tugging at one another due to their gravity?

Did God have to intervene from time to time to keep everything in its proper place? If so, why would a divine creator build such an imperfect system? Newton's scientific insight was extraordinary and his contributions to human knowledge among the greatest of all time, but he never found an adequate solution to this problem. It was not until the modern era that scientists realized that the solar system does not run eternally like clockwork with everything in its place. Eventually, the planets might stray far enough to collide with one another.

Newton's *Principia* included a method for finding the size, shape, and orientation of a comet's orbit given the comet's positions at three different points in time. Newton was spurred to develop this method by the appearance of bright comets in November and December of 1680 and a dispute with the Astronomer Royal, John Flamsteed, as to whether these comets were one and the same. Flamsteed maintained there was only a single comet that had disappeared and returned to view after passing close to the Sun. Newton initially believed that two comets were involved but subsequently changed his mind. By about 1684, he was convinced that comets follow curved paths through the solar system due to the Sun's gravity and that a single comet traveling on a parabolic orbit would explain both objects seen in 1680.

Edmond Halley played an important part in encouraging Newton to write his *Principia* and personally financed its publication by the Royal Society in London. Halley was particularly excited by the prospect of applying Newton's orbit finding method to comets seen in the past, driven by the suspicion that the same comet could return on multiple occasions. In 1695, he wrote to Newton, "I must entreat you to procure for me of Mr Flamsteed what he has observed of the Comett of 1682 particularly in the month of September, for I am more and more confirmed that we have seen the Comett now three times, since ye Yeare 1531." Applying Newton's ideas, Halley concluded that the comets seen in 1531, 1607, and 1682 were indeed reappearances of the same object. The slightly different intervals between the apparitions could be accounted for if Jupiter's gravity had perturbed the comet's orbit in between. Halley published his findings in 1705 in *Synopsis of the Astronomy of Comets* and predicted that the same comet would be seen again in late 1758 or early 1759. In 1758, the comet that came to bear his name reappeared on cue, 16 years after Halley's death. It was the first time anyone had predicted when a comet would appear and was a powerful vindication of Newton's theory of gravity.

THE MISSING PLANET

Between the beginning and the end of the 17th century, our picture of the solar system changed beyond all recognition. By 1700, it was widely accepted that the planets orbited around the Sun with their movements controlled by a universal law of gravity. The other planets were presumed to be worlds in their own right and even likely to be inhabited. Comets moved unhindered between the planets on extended orbits oriented at random, while the planets moved in roughly the same plane as one another. Jupiter had a family of four satellites and Saturn five. Saturn was surrounded by a system of rings although their nature was unknown. The solar system was 20 times larger than previously believed, and the stars were distant suns distributed throughout infinite space. As yet, astronomers did not consider that the solar system could have been different in the past and might be evolving. The timing and manner of the solar system's creation remained matters of theology.

One feature of the solar system proved particularly intriguing to astronomers of the day. The planets' orbits are not equally spaced. Rather, the gaps between Mars, Jupiter, and Saturn are vast compared with those separating the four inner planets. Oxford professor David Gregory set out the numbers simply in his 1702 textbook, which remained standard for many years:

> [S]upposing the distance of the Earth from the Sun to be divided into ten equal Parts, of these the distance of Mercury will be about four, of Venus seven, of Mars fifteen, of Jupiter fifty two, and that of Saturn ninety five.

In 1766, Johann Daniel Titius, an astronomy professor at Wittenberg in present-day Germany, noted that the distances followed a simple arithmetical relationship. His discovery was buried in a footnote added to a book he was translating, but it was spotted a few years later by Johann Elert Bode, who included it in his own book in 1772, the year he became director of the Berlin Observatory. The formula worked only if there was an additional planet between Mars and Jupiter orbiting roughly 2.8 AU from the Sun. Bode became convinced that such a planet existed, writing:

> This latter point appears to follow in particular from the remarkable relationship that the six known major planets follow in their distances from the Sun. Call the distance from the Sun to Saturn 100, then Mercury is separated from the Sun by 4 such parts; Venus 4 + 3 = 7; the Earth 4 + 6 = 10; and Mars 4 + 12 = 16. But now comes a gap in this very orderly progression. After Mars there follows a gap of 4 + 24 = 28 parts, where up to now no planet is seen. Can one believe that the Creator of the Universe has left this position empty? Certainly no. From here we come to the distance of Jupiter by 4 + 48 = 52 parts and finally to that of Saturn by 4 + 96 = 100.

The distances given by Gregory and Bode are only approximate, but the near coincidence between the predicted and actual distances is striking, and this was good enough for Bode. The relationship became widely known as Bode's law, although today it is more fairly called the Titius-Bode law. A decade after Bode's book was published, the formula appeared to be validated by the discovery of a new planet—but it wasn't the missing one. . . .

In 1766, the German musician William Herschel settled in the English city of Bath, where he had been appointed as an organist. In his spare time, Herschel indulged the other passion in his life—astronomy. Using superior reflecting telescopes he had built himself, Herschel spent many hours becoming familiar with the night sky and honing his observational skills. On the evening of March 13, 1781, his attention was focused on stars in the constellation Taurus, when he noticed an unusual object. He wrote in his observing notes, "In the quartile near zeta Tauri the lowest of two is a curious either Nebulous Star or perhaps a Comet." Four days later, he observed the object again and found that it was in a slightly different place.

The mysterious object had moved in such a short space of time that it had to lie nearby, which meant it was part of the solar system. Herschel assumed he had found a comet. Others, such as the Astronomer Royal, Nevil Maskelyne, were not so sure. Over the next few months, the new object failed to grow the fuzzy coma and tail characteristic of a comet, and mathematicians began to compute its orbit. By the summer it was clear that Herschel had discovered a new planet instead, the first person in recorded history to do so. Six years later, Herschel found two moons orbiting the planet, which were later named Titania and Oberon. The name of the planet itself remained a matter of dispute for some time. Herschel wanted to call the planet "George," after his patron King George III, but this suggestion was greeted with little enthusiasm by scientists working in other countries. Ultimately, it was Bode who suggested Uranus in keeping with the mythological names of the other planets.

The new planet orbited the Sun at almost twice the distance of Saturn. At a stroke, the known solar system had doubled in size and had acquired a significant new member. Uranus's average distance from the Sun is 19.2 AU (192 in Bode's units), and it escaped no one's attention that this distance was close to the value of 19.6 AU predicted by the Titius-Bode law. This was all the more reason to believe that another planet was lurking in the gap between Mars and Jupiter at a distance of about 2.8 AU from the Sun.

ASTEROIDS ENTER THE SCENE

One astronomer in particular was both convinced and utterly determined to find the missing planet by making a systematic search of the part of the sky through which all the known planets traveled. He was the Hungarian-born Baron Franz Xaver von Zach, astronomer to the Duke of Saxe-Gotha and director of the Seeburg Observatory near Gotha, in what is now Germany. After beginning a search on his own in 1787, he quickly concluded that the task was so great it would require a collaborative effort by several people. In September 1800, he managed to persuade five others to join him and to use their influence to enlist the support of even more astronomers at observatories around Europe, increasing the team to 24. However, Zach's plans were soon overtaken by events elsewhere.

At Palermo, in Sicily, Giuseppe Piazzi had recently set up Europe's southernmost observatory and equipped it with a state-of-the-art "vertical circle," a specialist telescope designed to measure positions very precisely. Piazzi would certainly have been one of the astronomers Zach wished to co-opt, but he was busy making observations for a new star catalog. On January 1, 1801, while recording the positions of stars in the constellation Taurus, Piazzi found a faint, eighth-magnitude object that was not in his reference catalog. Within days, it was clear that the object was moving with respect to the stars. Over the following month, he recorded its position on 24 nights until it became lost in twilight and was no longer visible after dark.

Publicly, Piazzi announced he had probably discovered a comet, but to a friend he confided his real belief that "it might be something better than a comet." Piazzi's initial calculations showed the new object was moving on a nearly circular path between Mars and Jupiter at a distance of 2.7 AU from the Sun. In April, having clearly established his claim to discovery, Piazzi shared his observations with other astronomers, including Bode. Piazzi called his discovery Ceres Ferdinandea for the patron goddess of Sicily and to flatter his royal sponsor, Prince Ferdinand. The Ferdinandea part was quickly dropped and today we know this object simply as Ceres.

A new body had turned up almost exactly where the Titius-Bode law predicted it should be, but problems soon arose. The first issue was whether Ceres would ever be seen again. In principle, Ceres would become visible in the night sky later in the year but Piazzi's limited observations were not enough to determine its future position with any certainty using existing techniques. Luckily, this problem was solved by the timely intervention of the brilliant mathematician, Karl Friedrich Gauss, who calculated a precise orbit placing Ceres at an average distance of 2.767 AU from the Sun. Using this orbit, Ceres was found again by Zach on December 31 and independently the next day by one of his five collaborators, Heinrich Olbers.

A more serious problem was that Ceres appeared to be very faint. It was immediately obvious that a major planet at this location should look brighter than Ceres did. That meant Ceres had to be small. By February 1802, William Herschel was informing the Royal Society that Ceres could be no bigger than five-eighths the size of the Moon, a mere 2,000 km (1,200 miles), based on its brightness. A few months later he revised his figure down to 260 km (160 miles). We now know this was a gross underestimate, the modern value for Ceres's diameter being 952 km (592 miles). But it was clear that Ceres is tiny compared to all the other planets.

Then, on March 28, Olbers discovered a second similar object. Ironically, Olbers was searching for Ceres at the time and spotted the new body in the same patch of sky. Olbers named his discovery Pallas, and Gauss soon computed its orbit. Pallas moves on an elliptical orbit, tilted at 34 degrees to the main plane of the solar system, with an average distance from the Sun of 2.770 AU. Not only is the orbit of Pallas much more elongated and inclined than that of any planets, but the orbits of Ceres and Pallas cross one another, making a collision possible.

Clearly, astronomers were dealing with a new class of objects that were neither major planets nor comets. William Herschel dubbed these objects "asteroids" due to their star-like appearance. Olbers immediately speculated that Ceres and Pallas were pieces of a larger planet that had disintegrated. If that were true, there were likely to be more fragments, and the appropriate places to look would be the points on the sky where the orbits of Ceres and Pallas appeared to cross. The hunt began, and a third asteroid, Juno, was discovered by Karl Harding in September 1804. Olbers found a fourth object, Vesta, in March 1807. At this point, the flow of discoveries dried up. For some time, astronomers continued to describe the four asteroids as "planets," although they were fully aware of their diminutive size. Textbooks and almanacs of the day included the asteroids along with the major planets in order of average distance from the Sun.

No new asteroids were found for several decades. Professional astronomers gradually stopped looking for them in order to pursue other more productive activities, and it fell to a dedicated amateur observer, Karl Ludwig Hencke, to continue the search. Hencke began looking in 1830 and showed extraordinary persistence despite finding nothing for 15 years. His patience was finally rewarded in 1845, when he discovered the fifth asteroid, now called Astraea. Two years later, he found asteroid number 6, Hebe.

With Hencke's discoveries, the drought ended and a steady stream of new finds ensued. The catalog of known asteroids has been growing ever since. As more asteroids were discovered, a clear picture began to emerge. Most lie in a broad, ring-shaped region between Mars and Jupiter. The distribution of their orbits and the variety in their composition mean that they cannot all be fragments of a single planet that exploded. The pace of discovery accelerated in the 1890s when photographic surveys began, and increased again in the 1980s and 1990s with the use of dedicated survey telescopes using charge coupled devices (CCDs). Since 1847, new asteroids have been discovered every year except 1945, and more than 300,000 are known today.

ROCKS IN SPACE

Even as the first asteroids were being discovered, a chance sequence of events showed that space must contain numerous smaller rocky objects as well. Folklore is rich with stories of stones falling to Earth from the sky. Scholars generally attributed these apparitions to a terrestrial origin, such as volcanoes, or dismissed the stories altogether as mere superstition. As Newton wrote in 1704, "[T]o make way for the regular and lasting motions of the planets and comets, it's necessary to empty the heavens of all matter, except perhaps some very thin vapours, steams or effluvia arising from the atmosphere of the Earth, planets and comets."

This view was widely shared by scientists in the 18th century, but attitudes were about to change.

In 1791, physicist Christoph Lichtenberg observed a bright fireball in the sky at Göttingen, in Germany. Fascinated by his account, fellow German physicist Ernst Chladni began investigating historical reports of other fireballs and stones falling from the sky. In 1794, he published a book in which he proposed that these rocks—what we now call meteorites—plunge to Earth from interplanetary space. Chladni's idea was initially ridiculed, but, in the very same year, many people witnessed a shower of meteorites falling near the Italian city of Siena. The following year, a farm laborer in Yorkshire, England, alerted by an explosion above him, was astonished to see a large stone emerge from the cloudy sky and land on the ground. The black stone, weighing about 20 kg (44 pounds), was still warm and smoking when he found it.

The event that really began to change scientific opinion occurred on April 26, 1803, near L'Aigle, in Normandy, France. Following three loud bangs, some 2,000 to 3,000 stones rained down on the fields near the town, prompting a full scientific investigation led by the French mineralogist Jean-Baptiste Biot. He concluded that the rocks had indeed come from space, which meant there must be many more such objects orbiting between the planets. Much later, in the 1960s, analysis of photographs of incoming fireballs associated with several meteorites revealed that they had come from the asteroid belt.

URANUS BEHAVING BADLY

When Hencke began his search for asteroids in the 1830s, looking in the part of the sky where the planets move, he may have had a second goal in mind. Uranus was behaving badly. Since its discovery in 1781, Uranus had been straying from the path it was expected to follow under the influence of the gravitational pull of the Sun and the other known planets. Several observers had recorded the position of Uranus before 1781 but had failed to recognize it as a planet rather than a star. When these "prediscovery" observations were included in the calculations, Uranus's deviation became even worse. An obvious explanation, widely discussed in astronomical circles by the 1830s, was that Uranus was being pulled off course by the gravity of a more distant undiscovered planet.

To find a major planet would bring international fame to any astronomer, professional or amateur. However, professionals generally regarded the task as hopeless without some theoretical prediction that would greatly narrow down the area to be searched. Predicting the new planet's location seemed a daunting mathematical problem, and it was dismissed as virtually impossible by eminent astronomers such as the British Astronomer Royal, George Biddell Airy, himself a talented mathematician.

Airy's opinion did not deter John Couch Adams, however. In 1841, while still an undergraduate student at St John's College, Cambridge, Adams recorded in his notebook, "Formed a design at the beginning of this week of investigating, as soon as possible after taking my degree, the irregularities of the motion of Uranus, which are not yet accounted for, in order to find whether they may be attributed to the action of an undiscovered planet beyond it." After graduating in 1843, Adams immediately set to work, assuming that the unknown planet's distance from the Sun was 38 AU as predicted by the Titius-Bode law. By September 1845, he believed he had pinned down the planet's position sufficiently for astronomers to find it. Unfortunately, the labors of an unknown young scholar failed to make an impression on Airy or the director of the Cambridge Observatory, James Challis. Months drifted by without action, and Adams himself was in no position to make the necessary observations.

Meanwhile, across the English Channel in Paris, Urbain Jean Joseph Leverrier started to tackle the problem in the summer of 1845, independently of Adams's work but using much the same methods. Leverrier announced his findings in June 1846 and sent a copy to Airy. His prediction for the new planet's position was almost identical to that of Adams, differing by less than a degree. This time Airy was stung into action, and he asked Challis to look for the planet as soon as possible. Astronomers at the Cambridge Observatory began a systematic search, but the request was not treated with particular urgency.

By contrast with the slow pace of events in England, Leverrier contacted an enthusiastic observer, Johann Gottfried Galle at the Berlin Observatory, who, as Leverrier had anticipated, lost no time beginning a search. On September 23, within an hour of pointing his telescope at Leverrier's predicted position, Galle found the planet we now call Neptune. A mere seventeen days later, English amateur William Lassell discovered Neptune's largest moon Triton.

The discovery of Neptune was a triumph for celestial mechanics. Galle found Neptune less than a degree away from Leverrier's predicted position and only 1.5 degrees from Adams's estimate. However, it soon became apparent that one of Leverrier's and Adams's basic assumptions was wrong. Like Uranus, several people had seen Neptune before it was discovered, including apparently Galileo. Combining these prediscovery observations with measurements made after 1846, astronomers found that Neptune actually lies 30.1 AU from the Sun, much closer than predicted by the Titius-Bode formula. With this failure, Bode's popular "law" was no longer taken seriously.

Despite the discovery of Neptune, minor discrepancies between Uranus's expected path and its actual motion became apparent as the 19th century wore on. At the start of the 20th century, two astronomers, Percival Lowell and William Henry Pickering, became convinced that something else was perturbing Uranus. The technique that had worked to spectacular effect in the case of Neptune might, they thought, reveal the presence of yet another planet. The charismatic Lowell, a Harvard mathematics graduate and wealthy businessman, built and equipped an observatory that bears his name in Flagstaff, Arizona. Lowell's initial interest was Mars, which he believed to be inhabited and crisscrossed by artificial canals. In about 1905, however, his attention turned to a possible "Planet X" lying beyond Neptune, and he began a search that continued until his death in 1916.

Lowell's calculations and searches were in vain, as were those of his rival Pickering. But in compliance with Lowell's wishes, the quest was renewed in 1925 with the commissioning of a telescope especially designed for planet hunting at Lowell Observatory. In 1929, observatory director Vesto Slipher hired a 22-year-old from Kansas, Clyde Tombaugh, to continue the search. Tombaugh was a quick learner and was soon systematically working his way around the sky, photographing each region on two different nights, separated by a few days. He then used an instrument called a "blink comparator" to examine each pair of photographs together. Switching back and forth between the two images made any object that had moved appear to jump. On February 18, 1930, a year after embarking on his task, Tombaugh found a tiny pinprick of light moving near the star Delta Geminorum. He recorded the "planet suspect" in his logbook and, after making some basic checks, went to Slipher's office. "I have found your Planet X," he said. The official announcement followed on March 13, and within weeks the new discovery had a name—Pluto.

But as with Ceres, the initial euphoria of discovery soon dissipated when it became apparent that Pluto didn't fit most people's notion of a major planet. By the summer of 1930, some 136 prediscovery observations of Pluto had been found, dating back to 1914, and these were used to calculate an orbit for the newly discovered body. It turned out to be an orbit unlike that of any of the known planets—highly elongated and tilted at 16 degrees to the main plane of the solar system. For part of its 248-year orbit, Pluto even crossed the path of Neptune, bringing it closer to the Sun than its neighbor. This was bizarre behavior.

Pluto was also very faint, even allowing for its great distance from the Sun. This meant Pluto was small, although just how small would not become apparent for another five decades. Its dwarf size and low mass were confirmed beyond doubt in 1978 after the discovery of Pluto's largest moon Charon. Within days, the motion of Charon allowed astronomers to measure Pluto's gravity and thus its size. Pluto is tiny, with a diameter of only about 2,000 km (1,200 miles)—the same as the north-south extent of the United States—while its mass is less than one-tenth of the Moon's. Even in the 1930s, it was clear that Pluto was too small to cause noticeable changes in Uranus's orbit. The mystery of the motion of Uranus was finally solved in the 1990s, when new calculations using better estimates for the mass and orbit of Neptune showed that Uranus is in fact precisely on course and there never was a need for a Planet X.

COMPLETING THE INVENTORY

From the 1960s onward, the pace of discovery in the solar system quickened with the advent of larger telescopes, sensitive light detectors, and space flight. Space missions in particular helped transform our view of the solar system, turning planets, moons, asteroids, and comets into real worlds with distinct personalities. We saw in Chapter 1 how the Mariner 4 mission to Mars began this transformation. This was followed by two Viking missions that landed on Mars's surface, sampled its rocks for the first time, and even looked for traces of life in the planet's soil without success. Numerous other missions to the Red Planet followed, culminating in a series of Mars rovers that have traveled across the planet's surface and, among their many discoveries, have found clear signs that liquid water has existed on the surface for extended periods in the past.

Space missions to the outer planets have produced a similar revolution in our understanding of these bodies. Among the most successful were the two Voyager missions that flew past all four giant planets between 1979 and 1989. The spacecraft returned close-up images of the giants' atmospheres, belts, and spots, discovered rings around Jupiter, found active volcanoes on Jupiter's moon Io and geysers on Neptune's moon Triton, and revealed extraordinarily complex structure within Saturn's rings that scientists are still working to understand today. These successes were followed by the Galileo and Cassini spacecraft, which began extended tours in orbit around Jupiter and Saturn in 1995 and 2004, respectively. As we will see in Chapter 12, Galileo became the first space mission to directly sample the atmosphere of a giant planet, and it also helped discover oceans in the interior of several of Jupiter's moons. Cassini revealed still more complexity within Saturn's rings and discovered geysers on Saturn's moon Enceladus, while the attached Huygens probe landed on Titan, discovering lakes of liquid hydrocarbons and giving us our first view beneath the moon's opaque atmosphere.

Spacecraft have provided by far the most detailed pictures of comets and asteroids we possess. In 1986, the Giotto mission to comet Halley took the first close-up pictures of a cometary nucleus, revealing an extremely dark, irregularly shaped body with several active jets of material escaping from its interior. The Galileo spacecraft flew by two rocky asteroids on its way to Jupiter in the early 1990s, allowing us to see their true shape and structure for the first time and discovering the first known moon orbiting an asteroid. The NEAR Shoemaker space mission gave us a detailed look at the dark, carbonaceous asteroid Mathilde in 1997, before going into orbit and ultimately landing on asteroid Eros in 2001.

In 2010, the Japanese Hayabusa mission became the first spacecraft to return samples from an asteroid, showing how rocky materials change over millions of years in the harsh environment of space.

Astronomers on Earth have continued to play an active role in exploring the solar system in recent decades, adding to the list of known asteroids, comets, and moons and discovering new types of object such as the Trojan asteroids that share orbits with Mars and Neptune. The number of known asteroids in the main asteroid belt and closer to Earth has grown dramatically in the past two decades thanks to dedicated telescopic surveys such as the Lincoln Near Earth Asteroid Research project. The 1990s saw the discovery of 1992 QB1 and hundreds of similar objects, which proved the existence of a large belt of icy bodies orbiting just beyond Neptune—the long hypothesized Kuiper belt. Pluto was no longer a lonely misfit but one of many thousands of trans-Neptunian objects populating the outer reaches of the solar system. We will return to this discovery, and what followed, in Chapter 14.

Today, after many centuries of discovery, we may finally have a comprehensive survey of the solar system, its architecture, and the great variety of objects it contains. In the next chapter we will see how our ever-improving picture of the solar system has shaped theories for its origin.

THREE

AN EVOLVING SOLAR SYSTEM

A CHANGING WORLD

Sixty-six years after James Cook traveled to Tahiti to observe the transit of Venus, the island received another scientific visitor on a historic voyage. "Crowds of men, women and children were collected on the memorable Point Venus ready to receive us with laughing, merry faces," wrote Charles Darwin of his arrival in 1835.

Four years earlier, when he was only 22, Darwin had been appointed as a naturalist on board the Royal Navy survey ship HMS *Beagle*. On a momentous five-year voyage around the world, Darwin carefully observed wildlife and collected specimens at several exotic locations, including Tierra del Fuego and the Galapagos Islands as well as Tahiti. These observations led Darwin to write one of the most influential books in the history of science, *On the Origin of Species by Means of Natural Selection*, published in 1859.

The central theme of Darwin's book is that all living species are descended from common ancestors. Each organism inherits a mixture of characteristics from its parents together with a few random mutations. Some descendants are better suited to their surroundings than others, and these descendants thrive while their ill-adapted cousins do not. Today we call this adaptation by natural selection "evolution," although surprisingly, Darwin hardly ever used the term himself.

Darwin suggested that new species arise and others become extinct naturally without divine intervention. As a respected geologist as well as a naturalist, he also believed that Earth changes gradually due to natural processes. Darwin wasn't the first person to assert that creatures evolve over long periods of time. His own grandfather, Erasmus Darwin, had suggested as much, and naturalist Alfred Russel Wallace developed his own theory of evolution independently of Darwin. However, Darwin's book was extremely popular and widely read. His writing helped establish a new way of thinking: the world and its creatures aren't static; they change slowly over time. For many people, however, the challenge was how to reconcile these evolutionary ideas with the account of creation and the great flood described in the Old Testament of the Bible.

At the time, the most widely known chronology of the world was James Ussher's *Annals of the World*, published in 1650. Ussher, who was Anglican Archbishop of Armagh (now in Northern Ireland), wrote this scholarly work after making a detailed study of the Bible and other ancient manuscripts. He concluded that Earth was created on October 23, 4004 BC. Other scholars, including Kepler and Newton, made comparable calculations and arrived at a similar age for the world, but Ussher's analysis was considered the most authoritative, and widely circulated English editions of the Bible referred to it in an annotation from 1700 onward.

A NEBULOUS IDEA BEGINS TO TAKE SHAPE

Ussher made two important assumptions. The first was that a chronology of the physical world could be worked out using documented human history alone. Ussher also assumed that the world—and by implication the whole universe—looks the same today as it did when it was created, the only significant changes having happened during the great flood recorded in the biblical book of Genesis. Before scientific theories about the origin and evolution of the solar system could be developed, these assumptions had to be abandoned. In this chapter, we will look at how our ideas about the beginning of the world have changed since Ussher, leaving the question of the timescales involved to the next chapter.

One person in particular helped to change the way we think about the origin of the solar system: Pierre-Simon de Laplace. Born in France in 1749, Laplace was regarded as an exceptional mathematician while still in his 20s, and he went on to earn a reputation as one of the most outstanding scientists of all time. When Laplace's ideas on the origin of the solar system appeared in print, they garnered considerable attention.

As part of a job he held briefly as a professor at the École Normal in Paris, Laplace gave a series of general lectures, and it was in one of these that he first outlined his "nebular hypothesis" for the origin of the solar system. Despite his mathematical prowess, Laplace presented his ideas in purely descriptive terms to make them accessible to a general audience. The lectures later formed the basis of a book, *Exposition of the System of the World*, which was published in 1796.

Laplace imagined that the planets began life as an extended cloud of gas slowly rotating about the Sun. Over time, the gas cloud cooled and contracted, rotating more rapidly as it did so and flattening into a disk. As the rotation rate increased, centrifugal force began to overpower the Sun's gravity, expelling rings of gas from the outer edge of the disk. Each of these rings later condensed into one of the planets we see today. Laplace suggested that the same process happened in miniature about each of the planets, forming families of satellites. Comets, with their highly elongated orbits, did not fit readily into the model, and Laplace deduced that these must be visitors from beyond the planetary system.

Laplace insisted that any viable theory for the solar system must account for the features that we see today. In this respect, the nebular hypothesis appeared to be a success. It explained why the planets all orbit the Sun in the same direction, and travel in nearly the same plane, and also why the planetary orbits are nearly circular.

Laplace's theory contrasted sharply with an earlier idea developed by the French naturalist Georges-Louis Leclerc, Comte de Buffon. In 1749, Buffon published a popular book suggesting that the planets were created 75,000 years ago when a comet collided with the Sun. This scenario at least had the merit that it invoked a single process rather than a series of unconnected events. However, Laplace and other scientists discounted the idea on the grounds that it did not fit the observed features of the solar system. Buffon's theory was also condemned by academic theologians of the day, and he soon retracted it to avoid trouble with the Church.

A few decades later, following the French Revolution, the political and religious situation was very different. The Church's power was greatly diminished and Laplace encountered few difficulties. Famously, when Napoleon asked Laplace why the Creator didn't feature in his treatise on celestial mechanics, Laplace is said to have replied, "Sir, I have no need of that hypothesis." Laplace meant that he saw no need for a higher power to intervene to maintain the stability of the solar system. This was a problem that had long vexed Newton, who had concluded that God must act from time to time to prevent the planets from straying due to their gravitational interactions. Laplace calculated that, rather than building up over time, these interactions cause only small oscillations in the planetary orbits. In 1788 he wrote, "Thus the system of the world only oscillates around a mean state from which it never departs except by a very small quantity . . . it enjoys a stability that can be destroyed only by foreign causes."

Laplace believed that the orbits of the planets were stable forever and that their past and future motions could be computed indefinitely. Some 70 years later, another French mathematical genius, Henri Poincaré, discovered that Laplace's theory applies only in certain special cases. Poincaré showed that the orbits of the planets are not necessarily stable after all, and the long-term evolution of the planets is highly sensitive to their exact starting conditions. In so doing, Poincaré laid many of the foundations for what we now call chaos theory. In reality, it is impossible to determine the motion of the planets far into the past or the future with certainty.

When Laplace published his *Exposition* in 1796, he was probably unaware that a different version of the nebular theory had already appeared 41 years earlier. Today, Immanuel Kant is known as a philosopher, but in his youth he studied physics and mathematics. One of Kant's earliest works was *Universal Natural History and Theory of the Heavens*, published in 1755, in which he described a model for the origin of the solar system.

In Kant's view, the solar system began as a cloud of material in space. Turbulent motions brought material together, allowing clumps to form, and these clumps attracted more material toward them. Denser clumps progressed more easily toward the Sun so that, when planets ultimately formed, the inner planets were denser than outer ones. Like Laplace, Kant presented his ideas in words and did not attempt to support his case with mathematical reasoning. Sadly for Kant, his publisher went bankrupt almost immediately, and the stock, including his book, was seized. Kant's work vanished almost without trace, and for many decades it was hardly known outside Kant's home city of Königsberg in Prussia.

Laplace never referred to Kant's work in the first edition of his *Exposition*, nor in the four revised editions that he produced in the 30 years before his death in 1827. However, Kant later found champions among his German compatriots. Chief among these was Hermann von Helmholtz, a highly influential physicist, philosopher, and physiologist.

On February 7, 1854, Helmholtz gave a public lecture in Königsberg, almost 50 years to the day since Kant's death, and nearly a century after Kant's *Universal Natural History* first appeared. Helmholtz's main theme was the conservation of energy, a principle he had publicized in 1847, drawing on the earlier work of others. No doubt he felt it was appropriate to honor Kant under the circumstances, and he told his audience:

> It was Kant who . . . seized the notion that the same attractive force of all ponderable matter which now supports the motion of the planets must also aforetime have been able to form from matter loosely scattered in space the planetary system. Afterwards, and independent of Kant, Laplace, the great author of the *Mécanique céleste* laid hold of the same thought and introduced it among astronomers.
>
> The commencement of our planetary system, including the Sun, must, according to this, be regarded as in an immense nebulous mass which filled the portion of space now occupied by our system far beyond the limits of Neptune, our most distant planet.

Helmholtz glossed over the many disparities between Kant's and Laplace's theories, concentrating instead on the common theme of a nebula, from which matter somehow came together to form the planets. As a result of Helmholtz's promotion, the nebular hypothesis came to be known as the "Kant-Laplace theory," despite the differences between the two men's ideas.

For many decades, the nebular hypothesis was the only scientific theory for the origin of the solar system. Some scientists criticized the theory for its lack of details and rigorous calculations, but on the whole it was regarded favorably. Laplace drew encouragement from new observations that appeared to support his idea. At that time, there was much speculation about mysterious cloud-like objects in space called nebulae. In 1811, William Herschel published a long paper describing his observations of some nebulae that each appeared to be made of shining fluid condensing around a central star. He suggested that these nebulae were stars in different stages of formation. Herschel never mentioned the nebular hypothesis in his paper, but Laplace seized on his observations and referred to them in the fifth edition of his *Exposition*. To Laplace, these nebulae looked like other solar systems in the making, just as his theory predicted.

THE NEBULAR HYPOTHESIS IN TROUBLE

By the 1840s, however, new observations of nebulae began to cast doubt on Laplace's ideas. William Parsons, Third Earl of Rosse, constructed the world's largest telescope in the grounds of his castle in the west of Ireland. The new telescope had a cumbersome design, but its huge 1.8-meter (72-inch) mirror allowed Rosse to see nebulae in greater detail than ever before (Figure 3.1). In 1845, Rosse reported that all the nebulae that he had looked at were actually composed of a huge number of faint stars. If Rosse was right, these were not nascent planetary systems but something else entirely.

Some scientists also had serious reservations about the nebular hypothesis on theoretical grounds. One was Adam Sedgwick, professor of geology at Cambridge University and mentor of the young Charles Darwin. In 1850, Sedgwick published a long list of problems with Laplace's theory, although he did not discount the nebular model altogether. Sedgwick raised a specter that continued to haunt researchers into the 20th century: how to explain the angular momentum (rotational inertia) of the Sun and the planets. One simple example illustrates the problem. Suppose Mercury formed from a ring of material expelled from a shrinking nebula. At that stage, the nebula must have been rotating once every 88 days—the time it takes Mercury to travel around the Sun today. As the nebula shrank further to form the Sun, it would have rotated ever more quickly due to its angular momentum. Today, the Sun should be rotating once every few days, when it actually takes almost a month to do so.

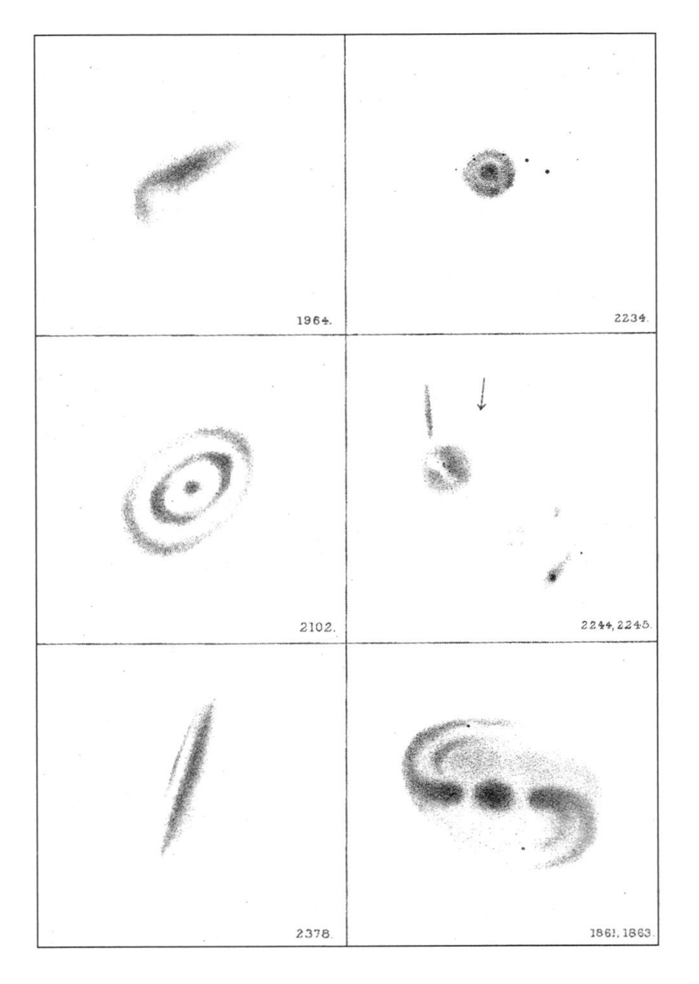

Figure 3.1. Drawings of nebulae made by William Parsons, 3rd Earl of Rosse, from *Observations of Nebulae and Clusters of Stars Made with the Six-foot and Three-foot Reflectors at Birr Castle from the Year 1848 up to the Year 1878*, published by Lord Rosse's son, Lawrence Parsons, 4th Earl of Rosse, in *Scientific Transactions of the Royal Dublin Society*, vol. 2, 1878. (J. Mitton)

In 1861, the French physicist Jaques Babinet raised the same objection, and the problem became more widely known. The Sun possesses more than 99 percent of the total mass of the solar system but only 2 percent of its rotational inertia; the motion of the planets takes up all the rest. In a contracting nebula, how could most of the mass end up in the center if it didn't take the angular momentum with it? Despite valiant attempts by others to modify Laplace's theory, this "angular momentum problem" was widely seen as a major flaw. Scientists identified other problems as well. The American mathematician Daniel Kirkwood showed that Laplace's disk would shed material continuously at its outer edge rather than forming discrete rings. Even if rings did form, Scottish physicist James Clerk Maxwell proved that a ring of material could not evolve into a single planet.

Laplace's theory was in serious trouble, but the nebular hypothesis still had some attractive features in the late 19th century. Radioactivity and nuclear reactions were unknown at the time, and it was widely believed that the Sun shines because it is gradually shrinking, releasing gravitational energy as it does so. The Sun must have been larger in the past and this fitted naturally with the idea that the Sun and the planets began as a diffuse cloud of material. Physicists had worked out the Sun's composition using features in its spectrum and found much in common with the composition of the planets, another plus for nebular theories. Astronomers were also discovering new types of nebulae that were clearly made of gas rather than stars, reviving the hope that some of these might be newly forming planetary systems.

On the other hand, several recently discovered moons did not fit into the nebular picture. Neptune's moon Triton moves backward around its planet; the satellites of Uranus travel at right angles to their planet's orbit; Mars's moon Phobos moves around its planet so quickly that it continually overtakes the planet's rotation. None of these bizarre objects could be explained by the nebular theory.

Overall, it seemed that the nebular hypothesis contained a grain of truth but that the details were clearly wrong. Was there an alternative? If the distribution of angular momentum in the solar system didn't arise from a condensing nebula, could the planets have formed as the result of some external event instead?

A CHANCE ENCOUNTER?

In December 1905, Edwin B. Frost, the director of Chicago University's Yerkes Observatory, received a communication from one of his colleagues:

> This is therefore to inform you that on and after January 1st proximo, the solar system will be run on the new hypothesis. It is not expected that the transition will be attended by any jar or other perceptible perturbation or that the change from a gaseous to a planetesimal feed will occasion any nausea. Everything is expected to work smoothly. . . . The planets will be allowed to rotate as fast or as slow as they please without regard to the speed of their satellites, and these will be permitted to go round forwards or backwards as they see fit without incurring the suspicion of being illegitimate members of the family. The inclination of the Sun's axis will not be regarded as a moral obliquity but merely as a frank confession that once on a time he flirted with a passing star.

The writer of this facetious letter was the flamboyant geology professor Thomas Chrowder Chamberlin, who had just published the second volume of a geology textbook containing a new model for the origin of the solar system.

One implication of Laplace's nebular theory is that Earth must have been hot and molten when it first formed. Chamberlin argued that this was implausible because Earth couldn't have held on to its atmosphere if it were true. So Chamberlin began to examine whether the planets could have formed by the accumulation of cold, solid particles instead. This idea had been introduced several decades earlier but received little support at the time. To proceed further, Chamberlin needed a scenario that would form these particles in the first place.

Chamberlin had seen photographs of spiral-shaped nebulae and was taken with the idea that they formed when stars passed close to one another. During such an encounter, each star's gravitational tug could have pulled a stream of gas from its neighbor, forming spiral arms around the stars. Perhaps the Sun suffered a similar encounter at one point after which the resulting material cooled and formed the building material for the planets. This scenario didn't involve a shrinking nebula, so it avoided the angular momentum problem that had plagued the nebular hypothesis.

Lacking a background in astronomy, Chamberlin sought help from a young astronomer, Forest Ray Moulton, who was just finishing his PhD at Chicago. The two men teamed up. Chamberlin focused on the details of a stellar encounter that could lead to the formation of the planets,

Figure 3.2. An early photograph of a "spiral nebula"—the Andromeda Galaxy—published in 1893. This collotype plate appeared in *A Selection of Photographs, Stars, Star-Clusters and Nebulae* by Isaac Roberts. Roberts thought this object was a gaseous spiral nebula around a central star. In the accompanying text, he states, "These photographs throw a strong light on the probable truth of the Nebular Hypothesis, for they show what appears to be the progressive evolution of a gigantic stellar system. Much additional evidence of a similar confirmatory character will also be seen on examination of other photographs of nebulae which are given in the pages following." (J. Mitton)

while Moulton used mathematical tools to demolish Laplace's theory and bolster Chamberlin's scenario. The pair published their results separately in 1905 and 1906, and the new hypothesis came to be known as the Chamberlin-Moulton theory.

Moulton calculated that another star must have passed about as close to the Sun as Jupiter is today, drawing out long spiral arms of gas in the process. Knots of dense material would have formed within the arms, and these would have acted as centers for the accumulation of material into planets. Presumably, the gas cooled rapidly, forming small solid particles orbiting around the Sun. Chamberlin dubbed these particles "planetesimals." Over time, planetesimals moving on nearby orbits collided and merged, aided by their mutual gravity, eventually forming planets.

Chamberlin and Moulton's scheme involved two parts: the formation of a spiral nebula around the Sun, and the aggregation of solid particles within the nebula to form the planets. Chamberlin later dropped any suggestion that the spiral nebulae seen by astronomers were growing planetary systems. He realized that these structures are far larger than the solar system, and by the 1920s it was clear that spiral nebulae are actually galaxies of stars entirely outside the Milky Way.

The popularity of the Chamberlin-Moulton theory peaked in about 1915, having gained most support in the United States. The theory's most ardent early detractor was the German astronomer Friedrich Nölke, who in 1908 criticized many of the details in Moulton's mathematical analysis. Nevertheless, whatever its weakness, the idea showed that there were alternatives to nebular theories. The prominent British scientists Harold Jeffreys and James Jeans were among early supporters of the encounter theory, and both came up with variations of their own. Each developed a modified scenario that overcame some of the difficulties with the Chamberlin-Moulton model, but these modifications raised as many problems as they solved.

Jeffreys later began to have doubts, pointing out that any stellar encounter close enough to produce a spiral nebula and form planets must be incredibly rare. The American astrophysicist Henry Norris Russell showed that any planets formed by a stellar encounter would orbit very close to the Sun, completely unlike the planetary system we see today.

Many scientists also viewed Chamberlin's planetesimal model with considerable skepticism.

It seemed that neither the nebular hypothesis nor the Chamberlin-Moulton scheme could be correct. In his 1935 book *The Solar System and Its Origin*, Russell bemoaned the situation:

> We are like a group of engineers trying to find a route from the mouth of a canyon to the plateau above. Explorations up the stream, no matter what branch they follow, lead only into box canyons, up which they can go no further. Landing on the plateau, by some flight of the imagination, they find "draws" which lead downward—but when followed turn aside and evidently tend to quite different outlets. But here our allegory breaks down. The canyon may well be impassable, affording no through route. But the solar system must have had an origin of some kind.

The fatal blow for the stellar encounter theory came in 1939. Astrophysicist Lyman Spitzer showed conclusively that any material torn from the Sun would be too hot to condense into planets and would rapidly dissipate into space instead. The stage was set for a return of the nebular hypothesis.

NEBULAR THEORY RESURRECTED

One of the first actors on the scene was the German physicist Carl Friedrich von Weizsäcker. In 1943 he addressed the nebular theory's main shortcoming: how to form a slowly rotating Sun in a rapidly rotating nebula. Von Weizsäcker argued that turbulent motions within the nebula could redistribute angular momentum so that the central regions rotated slowly. Many features of his model didn't stand up to scrutiny, but nebular theories were once again in vogue.

Other scientists began to work on the problem and found more viable solutions. Swedish physicist Hannes Alfvén and British astrophysicist Fred Hoyle discovered that magnetic interactions could act as a brake, slowing the Sun's rotation and transferring rotational inertia to the material that would form the planets. For the first time in more than a century, the angular momentum problem no longer seemed insurmountable.

While most scientists focused on the problem of the nebula's angular momentum, there remained the question of how the nebula's gas and dust could be transformed into planets. In abandoning the Chamberlin-Moulton theory, researchers had lost sight of the second innovative aspect of Chamberlin's work: the growth of planets from planetesimals.

Luckily, the planetesimal concept was alive and well in the Soviet Union. One person above all revitalized this productive line of research. He was the Russian physicist Viktor Safronov. Building on work by his colleague Otto Schmidt at the Moscow Institute of the Physics of the Earth, Safronov developed a complete model for the origin of the planets based on planetesimals. In 1969, Safronov published his ideas in a book in Russian, *Evolution of the Protoplanetary Cloud and Formation of the Earth and the Planets*. When an English translation appeared in 1972, it had an enormous impact. As we will see in Chapters 8 and 9, Safronov's work has stood the test of time and now forms the basis of our current ideas about how the planets formed.

FOUR

THE QUESTION OF TIMING

"Earth's Age: 4.6 billion years" announced *Chemical and Engineering News* in November 1953, describing a dramatic breakthrough in the quest to discover when our planet formed. Two weeks earlier, American geochemist Clair C. Patterson had announced his finding at a meeting of the Geological Society of America. Patterson told the gathered delegates that Earth is precisely 4.55 billion years old, give or take about 70 million years. A reporter for *Chemical and Engineering News* caught Patterson's presentation and wrote it up as a story. In this rather unconventional way, the world learned the true age of Earth for the first time.

It was lucky that a journalist recorded Patterson's actual words at the conference. Patterson and his collaborators were much more guarded when they published their work, saying only that Earth's age was "greater than 4 billion years." Patterson's reticence was understandable. Although we can examine Earth in much greater detail than any other body in the solar system, the key to calculating Earth's age lay in rocks from outer space. What Patterson actually did was measure the age of a meteorite. To work out when Earth formed, Patterson had to assume that Earth and the meteorite formed at about the same time. This seemed plausible, but he lacked proof.

Three years later, Patterson had what he needed, a vital piece of evidence that removed any doubts about his earlier announcement, as we will see. Other scientists soon confirmed his result. Patterson's work has stood the test of time. Half a century later, after some minor tweaks, scientists' best estimate is that Earth formed around 4.48 billion years ago.

READING THE COSMIC CLOCK

Patterson had finally answered a question that had confounded scholars and scientists for centuries. Theologians, historians, geologists, physicists, chemists, biologists, and astrophysicists had all tried to work out when the world began, each bringing their own perspective to the task. Occasionally, these different approaches led to substantial disagreements and even bitter disputes. As recently as the 19th century, scholars were still laboring to reconcile the wildly different conclusions reached by scientists and experts on religious scripture. In the late 19th and early 20th centuries, physicists and geologists clashed repeatedly over the ages of Earth and the Sun, each side convinced the other must be wrong.

The age of Earth is a fascinating question in itself. However, answering this question is also a key step to understanding the history of the solar system, how it was born, and how it fits into the larger story of the universe. The problem for scientists is how to construct a timeline for the distant past using only the world we see today.

EARLY ESTIMATES: INGENIOUS—BUT WRONG

One of the first attempts to estimate Earth's age based on its physical properties was made by Benoit de Maillet, an 18th-century French diplomat and amateur naturalist. De Maillet pointed out that seashells could be found in sediments on land in many places around the world. He reasoned that Earth's surface must have been entirely covered by water long ago, and that sea levels have been falling steadily ever since. By observing the sea level today and estimating how fast it fell, de Maillet thought he could work out how long Earth has existed.

We now know that de Maillet's main idea was flawed. Sea levels have remained roughly constant with minor fluctuations for billions of years, so de Maillet's calculation is meaningless. However, de Maillet correctly realized that long-term natural processes could provide a way to measure the age of Earth. His suggestion that Earth could be as old as 2 billion years was very daring and prescient for its time. Until the early

20th century, the general consensus among scientists was that Earth was certainly thousands of years old and probably millions, but not billions.

Sir George Darwin, son of Charles, and widely regarded as the founder of modern geophysics, was another who tackled the problem. Darwin developed a theory for the origin of the Moon in which the Moon broke away from the young Earth when it was rotating much more rapidly than today. Darwin calculated that Earth must have been spinning about once every two hours in order for the Moon to escape. Over time, gravitational interactions would have slowed Earth's rotation and caused the Moon's orbit to expand. In 1898, Darwin showed that it would take at least 56 million years to bring Earth and the Moon to their present state, and he argued that Earth must be roughly this old as well. Alas, Darwin's theory for the origin of the Moon hasn't stood up to modern scrutiny, and neither has his estimate for the age of Earth.

John Joly, a professor of geology in Dublin, tried another approach based on an idea originally proposed by Edmond Halley. Joly reasoned that rivers are continually scouring salt from rocks on land and transporting it to the oceans. This suggested to him that the sea should be growing saltier over time, providing a way to estimate the age of Earth. In 1899, just a year after George Darwin published his estimate, Joly announced that Earth is 89 million years old. In 1909, Joly revised his estimate upward to 150 million years. With hindsight, we now know that Joly's reasoning was wrong and that both his estimated ages are much too low. Chemical reactions on the ocean floor are actually removing salt as fast as rivers supply it, so Joly's calculation was no more valid than those of de Maillet or Darwin.

GEOLOGY VERSUS PHYSICS

In the early 19th century, geologists realized that they could work out the sequence of events that had shaped the landscape by studying the different layers of rock laid down over time. By estimating how quickly layers of sediments accumulated and the rate at which other layers eroded away, they could see that Earth must be very old indeed. Although the true age of Earth remained out of reach, it was becoming clear that geological timescales are much longer than the entire span of recorded human history.

These early geologists were strongly influenced by the doctrine of uniformitarianism, championed by Charles Darwin's friend and mentor Charles Lyell. Uniformitarians argued that processes operating in the past were the same as those that happened today, and that the pace of change was similarly slow. This point of view naturally implied that Earth must be very ancient, and geologists were reluctant to accept estimates from other disciplines that suggested Earth was young.

Darwin's theory of natural selection also required long timescales in order to allow species to evolve, and he sought to bolster these timescales using his geological expertise. Darwin calculated the rate of erosion in an area in southern England called the Weald, where weathering had cut through a domed rock structure to expose layers beneath the surface. He concluded that Earth must be at least 300 million years old, and included this calculation in the first edition of his book *On the Origin of Species* in 1859.

Meanwhile, physicists were puzzling over a fundamental theoretical problem that also constrained the age of the solar system: what keeps the Sun shining? The only explanation to gain acceptance in the 19th century, despite known inadequacies, was first proposed by Hermann von Helmholtz, whom we met in Chapter 3. In the same 1854 lecture in which he drew attention to Immanuel Kant's nebular theory of the origin of the solar system, Helmholtz argued that the Sun is slowly contracting and cooling over time. As gas within the Sun falls inward, it releases gravitational energy that is converted into heat and light. This energy source wouldn't last long though. Helmholtz estimated that the Sun had a lifetime of only about 22 million years, and this was surely the maximum age of Earth as well.

This line of reasoning found a powerful advocate in Sir William Thomson, who later became Lord Kelvin and was one of the most prominent physicists of the 19th century. Kelvin's groundbreaking work in thermodynamics prompted him to wonder how newly discovered laws governing heat and energy could be used to estimate the age of the Sun and Earth. At the time, physicists knew of no source of energy that could keep the Sun shining for the length of time envisioned by Darwin. In 1862, Kelvin launched an attack on the methods used by the geologists, which he sustained virtually until his death in 1907. Kelvin was particularly critical of the uniformitarian philosophy and the empirical nature of the evidence that geologists relied on. If the Sun is cooling over time, it must have been hotter in the past and so must have Earth. It defied common sense to believe that layers of sediment were being laid down on Earth at a constant rate if the planet was cooling down over time.

Kelvin preferred to rely on the known physical laws of nature and the conclusions that could be drawn from them. He conceded that the Sun might have taken as much as 500 million years to cool to its present state, but it could also have taken as little as 20 million years. Later, Kelvin calculated the cooling rate for Earth assuming that it was hot and molten when it first formed. He arrived at an age of 20 to 400 million years, similar to his estimate for the age of the Sun. This spurred him to renew his assault on uniformitarianism. In subsequent years, Kelvin's position hardened in the direction of the lower figure of 20 million years. In 1893, the American geologist Clarence King refined Kelvin's method and arrived at an age of 24 million years, a result that Kelvin endorsed.

Lord Kelvin was held in such high regard that few people dared to challenge him. Physics was widely seen as a more fundamental science than geology, and Kelvin's views were regarded as authoritative for three decades after he first entered the debate. Unfortunately, Kelvin and King greatly underestimated the complexity of Earth's thermal history, which meant that their elegant method for calculating Earth's age was seriously flawed.

Some geologists continued to pursue sedimentation as a way to find the age of Earth. One of the most thorough studies was published in 1893 by the American geologist Charles Walcott, who succeeded Clarence King as director of the U.S. Geological Survey the following year. Walcott concluded that Earth is 55 million years old, somewhat older than Kelvin's estimate but far below the figure accepted today. With hindsight, we can see that Walcott's method was no better than any of the others tried in the 19th century. Although Kelvin was unable to see the flaws in his own approach, he was correct that rates of sediment deposition and erosion have varied widely over time. A more fundamental shortcoming is that the sedimentary record is far from complete and is impossible to analyze for the first nine-tenths of Earth's history.

RADIOACTIVITY CHANGES EVERYTHING

One geologist refused to be intimidated by Kelvin's onslaught. He was the outspoken Chicago geology professor Thomas Chrowder Chamberlin, who developed his own theory for the origin of the solar system, as we saw in Chapter 3. In 1899, Chamberlin wrote a powerful rebuttal of Kelvin's methods and assumptions in the prestigious journal *Science*. Chamberlin's reasoning was uncannily prophetic:

> What the internal constitution of the atoms may be is yet an open question. It is not improbable that they are complex organizations and the seats of enormous energies. Certainly, no careful chemist would affirm either that the atoms are really elementary or that there may not be locked up in them energies of the first order of magnitude. . . . Nor would he probably feel prepared to affirm or deny that the extraordinary conditions which reside in the center of the sun may not set free a portion of this energy.

Chamberlin realized that a dramatic revolution in physics was just beginning, one that would have far-reaching implications. In 1896, the French physicist Henri Becquerel had discovered that uranium salts are radioactive. By 1898, Marie and Pierre Curie had discovered two new radioactive elements, polonium and radium, and Marie Curie had shown that atoms of these elements release powerful ionizing radiation. It soon became clear that radioactive materials generate substantial amounts of heat in Earth's interior, rendering Kelvin's cooling calculation useless. More important, radioactivity would provide a new and robust way to measure the age of ancient rocks, making all other geological dating methods obsolete.

In a series of experiments between 1901 and 1903, the New Zealand–born physicist Ernest Rutherford and British chemist Frederick Soddy discovered that radioactive elements could spontaneously transmute into new ones, releasing radiation as they did so. They identified three distinct types of radiation, labeled alpha, beta, and gamma. Alpha radiation consists of the nuclei of helium atoms; beta rays are electrons; and gamma rays are an energetic form of electromagnetic radiation similar to X-rays. Rutherford and Soddy found that the level of activity in a radioactive sample was proportional to the number of radioactive atoms that are present. This means that the activity of a particular element falls by half in a characteristic period of time called the half-life, which varies from one element to another (Figure 4.1).

Rutherford recognized that radioactivity represents a natural clock built into radioactive substances. He quickly saw the potential to use radioactive elements with very long half-lives to measure the ages of rocks. In a lecture he gave in 1904, Rutherford illustrated the point by showing that two samples containing uranium must have ages of about 500 million years, based on the amount of helium they contained. As the uranium atoms in the rock decayed, they released alpha radiation that

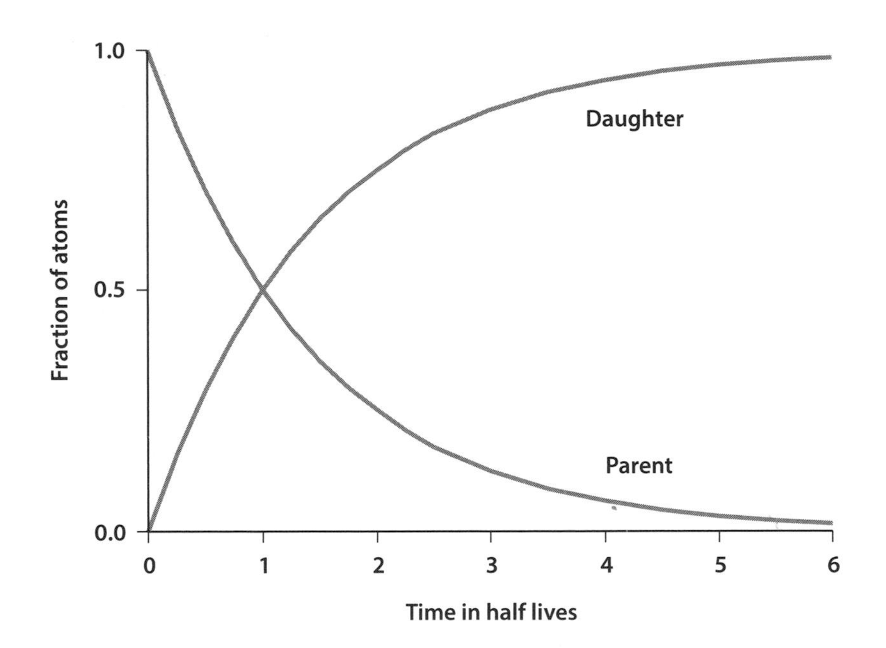

Figure 4.1. The half-life of a radioactive isotope. The number of atoms of a radioactive "parent" isotope halve over each successive period of one half-life. At the same time, the number of atoms of the "daughter" isotope produced by the decay rises.

became trapped inside the rocks as helium gas. As Rutherford pointed out, it was very likely that some helium had escaped, so the ages he measured were probably an underestimate.

The following year, the American physicist and chemist Bertram Boltwood discovered that uranium minerals always contain lead as well as helium. He deduced that lead is the final material to form when uranium decays into other radioactive elements. Unlike helium, lead cannot escape easily from rocks, so measuring the amount of lead in a sample should provide a more reliable age than measuring helium. Boltwood began to apply his new method to 43 rock samples collected from ten different locations. In 1907, he published his findings: the rocks had formed between 410 and 2,200 million years ago. Earth itself must be at least this old if not older.

In the 1920s and '30s, radiometric measurements of rocks were firmly pushing Earth's age upward. Radiometric dating methods were still imperfect and subject to a good deal of uncertainty, but scientists were routinely measuring Earth's age in billions of years rather than millions. A few estimates went as high as 3.5 billion years, while the generally accepted figure hovered between 1.6 and 2.0 billion years. Opinions on this matter were surely influenced to some extent by some astonishing new data obtained by astronomer Edwin Hubble, who for the first time had estimated the age of the entire universe.

HUBBLE AND THE AGE OF THE UNIVERSE

At the start of the 20th century, astronomers knew virtually nothing about the universe beyond our Milky Way galaxy. Stars, star clusters, and nebulae of all kinds were thought to belong to the Milky Way. This viewpoint began to change in 1917 when Vesto Slipher showed that 25 spiral nebulae are moving away from us at extraordinarily high speeds. Slipher's discovery disposed of the notion that spiral nebulae are planetary systems in the making once and for all. It also fueled a growing debate about whether these spirals lie within the Milky Way or are separate galaxies in their own right. In 1925, Hubble settled the issue. He found that spiral nebulae in the constellations Andromeda and Triangulum lie at immense distances. These nebulae could not possibly be part of the Milky Way and must be separate galaxies.

Four years later, Hubble had enough data to show that the great majority of galaxies in the universe are receding from us. He also found that each galaxy's recession speed is roughly proportional to its distance, a relationship we now know as Hubble's law. At about the same time, mathematicians discovered that Einstein's revolutionary theory of gravity—his General Theory of Relativity—naturally predicted that the universe could be expanding, with galaxies moving apart from one another, consistent with Hubble's observations.

By measuring how fast the universe is expanding, astronomers could work out how old it is, winding the clock backward to a time when the material in the universe was crammed together to an infinitely high density. This gave a figure of 1.8 billion years. Scientists knew that this estimate and calculations based on radiometric dating were both somewhat uncertain, but it was comforting that the two numbers were broadly similar.

By the 1940s, however, geologists' best estimate for the age of Earth had risen to more than 3 billion years, while Hubble's age for the universe remained stubbornly at around 2 billion. In 1952, the reason for this discrepancy became clear when astronomers realized that Hubble's technique for measuring the distances to other galaxies included a step based on erroneous data. Astronomers quickly fixed the problem, and the age of the universe increased by a factor of three overnight. The paradox of an Earth that appeared to be older than the universe was solved, but scientists still had much work to do before they would get an accurate age for either.

HOW RADIOACTIVE TIMERS WORK

Radiometric dating has become such an important tool in the quest to determine the history of the solar system that it is worth taking a closer look at how it works and why scientists place so much confidence in its results.

Every atom has a nucleus at its center composed of two types of particle: positively charged protons and neutral neutrons. A swarm of negatively charged electrons surrounds the nucleus, and these electrons give an atom its chemical properties. Atoms are neutral overall, so the number of electrons matches the number of protons, also known as an atom's atomic number. A chemical element can exist in multiple forms called isotopes, each having the same number of protons per atom, but a different number of neutrons. This means that different isotopes of carbon, for example, share the same atomic number but have different atomic masses. Every isotope can be described uniquely by naming which element it is and giving its atomic mass, which is a whole number. Carbon-12, for example, refers to atoms of carbon that have a nucleus containing 12 neutrons and protons in total.

Radioactive atoms have an unstable nucleus. At some point in time, each unstable "parent" nucleus emits a particle, or gamma radiation, or both, and decays into a "daughter" nucleus. When a nucleus emits a particle, the atom changes from one chemical element into another. It is impossible to predict when an individual atom will decay. However, a large collection of atoms decays at a predictable rate that depends only on the isotope involved. On average, half the atoms in a large sample will decay over the period of one half-life. After two half-lives, three-quarters of the sample will have decayed. Seven-eighths will be gone after three half-lives, and so on.

Scientists have measured the half-lives of many radioactive isotopes in the laboratory, some of which can be used like a stopwatch to measure the ages of rocks. When a new rock forms, in a lava flow for example, its minerals are typically hot, and atoms can move about freely within it. Once a mineral cools, its atoms become trapped in place and stay there, even though they may subsequently decay into a different element. So, if atoms of a suitable radioactive isotope are caught within a rock as it forms, it has a built-in timer. "Suitable" isotopes have a half-life measured in millions of years and are present in quantities large enough to be measurable with reasonable accuracy. Once the atoms are trapped, the mineral's composition is fixed and the radiometric clock begins to tick.

To give an example, suppose you have a rock that contains the radioactive isotope rubidium-87 and a measurable amount of its daughter isotope strontium-87. If no strontium-87 was present when the rock first formed, it is straightforward to calculate how long the rock has existed by measuring how much strontium-87 it contains today. However, real life is seldom this simple. In most cases some strontium-87 would have been present when the rock first formed. The sample may also have lost some of its rubidium or strontium, or may have been contaminated. Fortunately, there are ingenious ways to overcome these difficulties.

The first step is to measure the concentrations of the parent and daughter isotopes relative to an isotope of the daughter product that isn't involved in radioactive decay. For example, rubidium-87 and strontium-87 can be compared with strontium-86, which is stable and not part of any decay process. The second trick is to examine several minerals from the same rock sample. Different minerals have different crystal structures. In our example, some minerals preferred to incorporate rubidium when they first formed, while other minerals began with more strontium. The minerals started out with different *total* amounts of strontium, but the *relative* amounts of strontium-86 and strontium-87 would have been the same since the processes that form minerals generally don't differentiate between isotopes of the same element.

The age of the rock can be found by plotting a graph that shows the amounts of rubidium-87 and strontium-87 relative to strontium-86 for each mineral (Figure 4.2). Minerals that originally contained large amounts of rubidium-87 will contain a lot of strontium-87 today. Minerals that started out with no rubidium-87 will contain only as much strontium-87 as they had initially. This means that the points on the graph will lie along a single line, which is called an isochron because it connects minerals of the same age. The steeper the isochron, the older the rock is.

Not all rocks can be dated this way. If a rock is strongly heated after it first forms, its radiometric clock can be disturbed as atoms become free to move around again. In this case, different minerals within the rock will no longer fall along an isochron, and it may be impossible to determine the rock's age reliably. Sedimentary rocks are also problematic. These are composed of mineral fragments eroded from older rocks. While the individual mineral crystals can be dated, it is much harder to say when these pieces came together to form a sedimentary rock. For

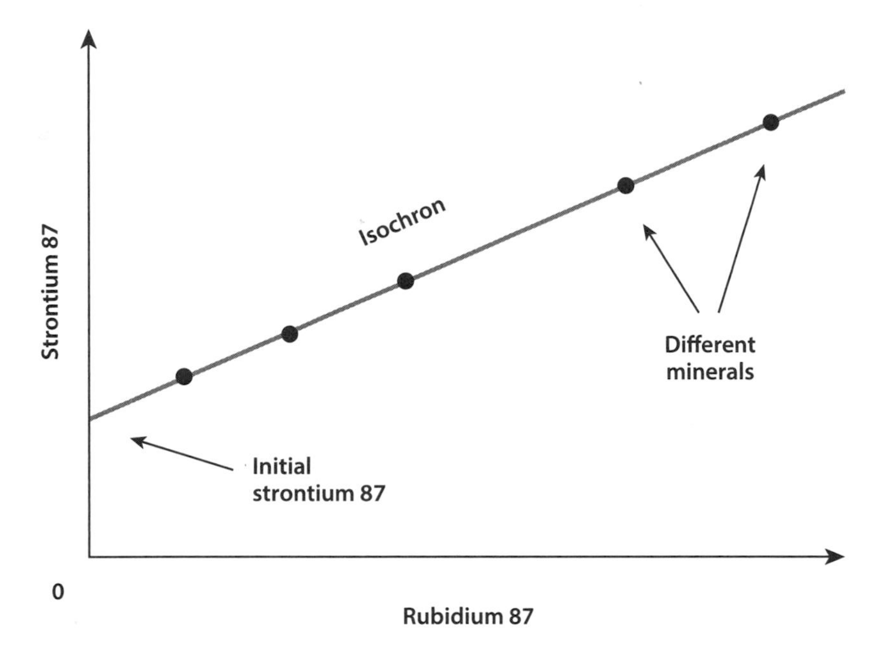

Figure 4.2. A schematic isochron. In this example, the amounts of radioactive rubidium-87 and strontium-87 have been measured relative to stable strontium-86 in five different minerals contained in one rock sample. Plotting the measurements on a graph reveals the age of the rock (see text). The steeper the isochron, the older the rock.

these reasons, radiometric dating results have to be interprcted in light of other clues within rocks, and it pays to select samples carefully.

This kind of radiometric dating can tell us the ages of individual rocks on Earth, but not the age of the planet itself. To find out when Earth formed, we would need rock samples from the earliest days of our planet's history. Unfortunately, Earth is a geologically active world, and its rocks are continually being destroyed and reprocessed. The oldest known rocks are thought to be significantly younger than Earth itself.

In the 1940s, three scientists found a solution to this problem, each operating independently and unaware of the others' work. These were the Russian E. K. Gerling, British geologist Arthur Holmes, and the German Fritz G. Houtermans. All three researchers exploited the fact that uranium has two naturally occurring radioactive isotopes, each of which decays ultimately into lead. Uranium-235 transmutes into lead-207, while uranium-238 becomes lead-206. Usefully, lead also has another isotope, lead-204, which is not radioactive or the product of radioactive decay. The two uranium isotopes have different half-lives, and this means that the relative amount of lead-207 and lead-206 on Earth changes over time. As long as we know the initial ratio of these isotopes and the mixture today, we can work out Earth's age regardless of how many times its rocks have melted and reformed.

The methods used by Gerling, Holmes, and Houtermans differed somewhat, but each tried to measure the mixture of the three lead isotopes in modern rocks and combine it with his best estimate for the initial ratio. Using somewhat different assumptions, Holmes and Houtermans arrived at an age of about 3 billion years for Earth, and Gerling nearer 4 billion. To do any better than this, scientists would have to get an accurate value for the initial mix of lead isotopes when Earth first formed.

METEORITES HOLD THE KEY

This is where Clair Patterson enters our story. Houtermans suggested that the initial mix of lead isotopes could be found by examining iron meteorites. As we will see in Chapter 5, meteorites come from asteroids that have remained largely unchanged since their early days of the solar system, so they should preserve a record of the solar system's initial composition. Patterson took up Houtermans's idea and measured the lead isotopes in a sample taken from the Canyon Diablo meteorite, which crash-landed in Arizona some 60,000 years ago. The amount of uranium in this sample was so small that its lead isotopes must have stayed almost constant since the meteorite's parent asteroid formed. Assuming this mixture really was primordial, Patterson could now calculate the age of the asteroid. In 1953, he published his findings: the asteroid was 4.55 billion years old. Soon afterward, Houtermans made a similar measurement and arrived at an almost identical conclusion.

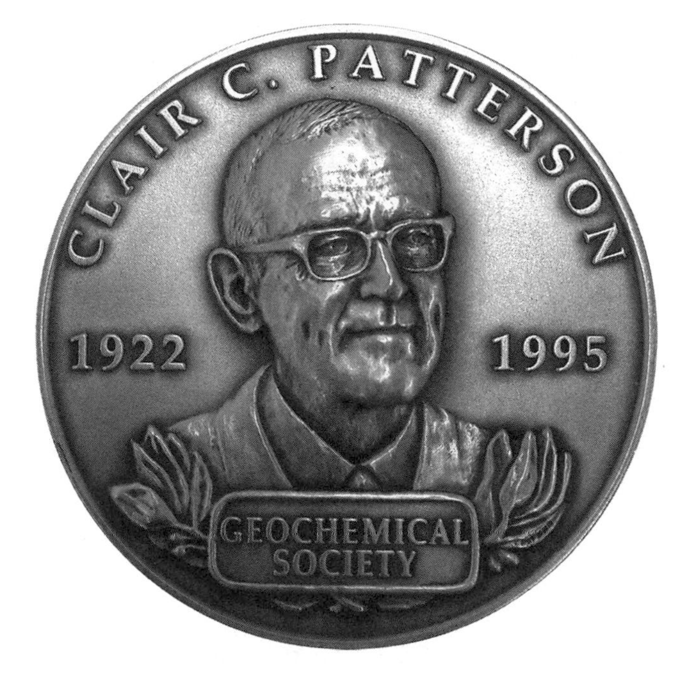

Figure 4.3. The Clair C. Patterson Award medal. The award is made annually by the Geochemical Society for a recent innovative breakthrough in environmental geochemistry of fundamental significance, published in a peer-reviewed journal. (Courtesy the Geochemical Society)

This left one vital question: did Earth and the asteroid have the same lead ratios to begin with? To test this, Patterson examined the lead found in deep-sea sediments from the floor of the Pacific Ocean, arguing that the mixture of lead isotopes in these sediments should be similar to that on Earth as a whole. The results fell precisely on the same isochron as data from five separate meteorites. This meant that Earth and the parent asteroids of these meteorites had all formed from a common reservoir of material with the same mixture of lead isotopes. The isochron gave an age of 4.55 billion years. This really was the age of Earth.

Since Patterson's work, other scientists have improved and refined radiometric dating methods. Researchers have measured the ages of a huge number of samples, including rocks from the Moon, Mars, and asteroids, as well as terrestrial rocks. We now know that every continent on Earth contains some rocks that are more than 3.5 billion years old. The oldest terrestrial samples of all are tiny zircon crystals embedded in sedimentary rocks in Western Australia, with ages between 3.5 and 4.4 billion years. These crystals are all that remain of the earliest rocks that formed on Earth, as we will see in Chapter 11. Rocks from the Moon tend to have ages between 3.2 and 4.0 billion years, with the very oldest being 4.45 billion years old. Rocks from the surface of Mars have a range of ages up to 4.5 billion years. The oldest samples of all come from meteorites. The majority of these ancient rocks formed between 4.4 and 4.6 billion years ago.

The fact that samples from different objects in the solar system have a range of ages should not be too surprising. Planets, moons, and asteroids must have taken some time to grow, and many of these objects evolved further after they formed. Precise radiometric dating now allows us to say more than just when the solar system formed. Instead, we can get a detailed timeline for many events early in its history.

The decay products of short-lived radioactive isotopes that no longer exist today can also provide important clues in this detective work. Their presence in meteorites tells us that the solar system formed from material that contained several radioactive isotopes with half-lives of less than 100 million years. The most plausible explanation is that these were created in a nearby supernova explosion shortly before the solar system formed. A good example is magnesium-26, a product of the decay of aluminum-26, which has a half-life of 700,000 years. The amount of magnesium-26 found in minerals that include magnesium varies from one meteorite to another. This strongly suggests that the parent asteroids of these meteorites formed at different times as the amount of aluminum-26 diminished.

We now know that the first solid materials to form in the solar system did so 4.566 billion years ago. The parent asteroids of most meteorites formed within the next few million years after this, with different asteroids forming at different times. Mars also formed at an early stage, certainly no more than 20 million years after the start of the solar system and probably sooner. Earth took significantly longer to grow, and the

Moon did not appear on the scene until at least 60 million years after the solar system began.

Clearly, radiometric dating is an incredibly useful tool for measuring the history of the solar system and its members. However, we are fortunate in having an independent way to check some of these numbers by measuring the ages of stars, including our Sun.

DATING THE SUN

Stars are essentially huge spheres of hot plasma and relatively simple objects to understand. Astronomers can work out the conditions deep inside a star using basic laws of physics and the known properties of matter at various temperatures and pressures. If a star is stable, the tendency of the plasma to expand outward is exactly balanced by the weight of material bearing down from above. This tells us how densely matter is packed within a star. In each layer of plasma, heat from the interior can move toward the surface either in the form of radiation or by convection—the continual rise of hot plumes of plasma through cooler material. Using this information, we can calculate the temperature in different layers within the star.

The British astrophysicist Arthur Stanley Eddington laid the foundations of this approach to stellar astrophysics in the early decades of the 20th century, proving among other things that the luminosity of a star is related directly to its mass. Eddington understood that the temperature in the center of a star must be measured in millions of degrees, and he believed that nuclear reactions provided the energy needed to maintain these temperatures. In the 1930s, nuclear physicists showed beyond doubt that the Sun draws its power from the fusion of hydrogen nuclei to form helium. The supply of nuclear fuel available inside a star is limited, however. The more massive a star is, the brighter it burns, and the sooner it will run out of fuel. A star like the Sun has a life span of roughly 10 billion years, but a star that is 10 times more massive lasts for only a few million years.

Although a star's life expectancy is predetermined by its mass, measuring the current age of a particular star is surprisingly difficult because its external appearance hardly changes during 90 percent of its lifetime. Astronomers describe stars in this long midphase of their life cycle as main-sequence stars. As a star's time on the main sequence draws to a close, it begins to expand and grow brighter—the star becomes a giant. Stellar evolution models tell us when this will happen to stars with different masses. If we know that a star is on the cusp of changing from the main sequence to a giant, we can deduce its age immediately.

However, it can be hard to judge whether an individual star is currently at this transition point. It becomes much easier to tell when looking at a group of stars with roughly the same age. Stars form in clusters, and stars in the same cluster tend to have a wide range of masses. Later on, when the more massive members of a cluster have evolved into giants, there will be a pivotal mass below which stars in the cluster still belong to the main sequence, and above which all the stars are giants. By pinpointing this mass, astronomers can calculate the age of the cluster. Finding the ages of star clusters has been extremely useful for understanding the history of star formation and the evolution of our galaxy as a whole. For example, it has revealed that very old stars contain much smaller amounts of heavy elements than do young stars. This has important implications for planet formation, as we will see in Chapter 7. Unfortunately, the Sun doesn't belong to a cluster, so astronomers had to find another way to measure how old it is.

The breakthrough came in 1960, when Caltech professor Robert Leighton discovered that the Sun is oscillating continuously, rather like the ringing of a giant bell. These oscillations cause gas at the Sun's surface to rise and fall over time in a regular pattern. The study of these oscillations, called helioseismology, turns out to be a powerful tool for measuring the physical conditions inside the Sun. For example, astronomers looking at one face of the Sun can tell if there are dark sunspots on the opposite face that would otherwise be undetectable.

The way these oscillations propagate through the Sun depends on the relative amount of hydrogen and helium in its interior. Astronomers have a good idea how much of each element was present in the material that initially formed the Sun, and the amount of energy produced by the Sun tells us how fast it is converting hydrogen into helium. Putting these together with the helioseismology data yields the Sun's age. An estimate published in 2011 suggests the Sun first became a main-sequence star 4.60 billion years ago, give or take 0.04 billion years. This is almost the same as the age of the oldest meteorites found by radiometric dating. As we saw earlier, having two independent measurements that tally does not necessarily mean that they are both right. However, unlike the methods employed by 19th-century scientists, both radiometric dating and helioseismology are firmly based on accurate observations and soundly tested theory. We can be confident that we have found the true age of the solar system at last.

THE AGE OF THE UNIVERSE REVISITED

One question remains: how does the age of the solar system compare to that of the universe as a whole? For decades, astronomers had tried to answer this question by refining Hubble's method to measure the expansion of the universe, and this was one of the key objectives for the Hubble Space Telescope. By measuring the distances and recession speeds of galaxies across the universe, astronomers could estimate the age of the universe using a theoretical cosmological model, together with other factors such as the density of matter and energy in the universe.

The most challenging aspect of this work is measuring the distances to very remote objects accurately. Since the 1980s, astronomers have measured how far away other galaxies are by observing stars within these galaxies that explode as supernovae. These explosions provide an excellent way to gauge distance since they are visible from very far away and their intrinsic brightness is always the same to within about 10 percent. Using these measurements, astronomers now have a much better idea of the scale and age of the universe.

In the 1990s, cosmologists came up with a new and independent way to determine the age of the universe using the cosmic microwave background. Discovered in 1964, this faint radiation travels through space in all directions, a relic from the Big Bang that formed the universe. Initially, scientists thought the background radiation would look the same in every direction. However, in 1992, a space mission called the Cosmic Background Explorer (COBE) discovered small variations in different parts of the sky. Astronomers realized that the scale of these fluctuations should depend on the age of the universe and how it has evolved over time. In 2001, a more advanced observatory, the Wilkinson Microwave Anisotropy Probe (WMAP), was launched to measure these fluctuations precisely. The new measurements, combined with the supernova data and other results, show that the universe is 13.75 billion years old, with an uncertainty of only 0.13 billion years.

One of the oldest known stars in the Milky Way has an age of 13.82 billion years, almost as old as the universe itself. The spectrum of this unusual star shows that it contains relatively large amounts of the radioactive elements uranium and thorium. This has made it possible to measure its age by the same kind of radiometric dating methods used on rocks in the solar system. The same dating method can also be applied to the Sun itself and has produced the same age as helioseismology and the radiometric dating of meteorites. At a mere 4.5 billion years old, the Sun and its planetary system are relatively new arrivals in the universe.

Thanks to a variety of techniques, we now have a reliable chronology of the major events in the history of the universe, the shaping of our galaxy, and the early evolution of the solar system. Meteorites in particular have played a central role in unraveling the history of the solar system. These invaluable samples of space rock are the subject of the next chapter.

FIVE

METEORITES

A DRAMATIC ENTRANCE

Just before midnight on a fall evening in 1992, a brilliant ball of fire appeared in the skies above the United States. Brighter than the full Moon, it was seen by hundreds of people as it traveled northeast across West Virginia, Maryland, Pennsylvania, and New Jersey. Distinctly green in color, and accompanied by sharp crackling sounds, the ball of light was filmed by a dozen onlookers, including several spectators at a Friday-night football game. Some of these films show the fireball separating into multiple parts, with at least 70 pieces visible at one point. The fireball finally vanished over New York State. The entire spectacle had lasted only 40 seconds.

Shortly after the fireball disappeared, Michelle Knapp, a resident of Peekskill, New York, was startled by a loud crash outside her house. Heading outside to investigate, she discovered that the Chevrolet Malibu parked next to her house had been turned into a wreck. The rear end of the car was crushed, and something had punched a large hole through one corner of the trunk. Lying next to the car was a basketball-sized lump of rock that felt warm to the touch. Gas was leaking from the car's tank, so the fire department was summoned along with the police. Officers initially suspected that vandals had damaged the car, even going so far as to impound the rock as a suspicious object. However, it soon became obvious that the rock's arrival and the car's destruction were directly linked to the fireball that had been seen by hundreds of people only moments earlier.

The events in Peekskill and elsewhere on that Friday evening are a dramatic example of a meteorite fall. The stone that hit the car in Peekskill was originally part of a beach-ball-sized boulder traveling through space. The boulder approached Earth at 15 km per second (34,000 miles per hour). As it slammed into the atmosphere, friction with the air generated tremendous amounts of heat, causing the outer layers of the boulder to glow white-hot and evaporate away. Far below, witnesses on the ground saw the glowing lump of rock and gas as a brilliant ball of fire, descending through the night sky.

Countless small rocks from space encounter Earth each day, burning up completely in the atmosphere as meteors. Most meteors are tiny, no larger than a grain of sand. These disappear in the upper atmosphere in a matter of seconds. The boulder at the center of the 1992 fireball was much larger—too large to burn up entirely—and it plowed straight through the upper atmosphere, continuing toward the ground. As the boulder reached the lower atmosphere, where the air is densest, the wind resistance became strong enough to tear the boulder to pieces. The rock's disintegration caused the crackling sounds heard on the ground and produced a swarm of smaller fireballs heading in the same direction. At least one piece of the boulder survived to reach the ground, where it became a meteorite—a lump of rock that literally fell from the sky.

WHERE DO METEORITES COME FROM?

In a few cases, such as Peekskill, astronomers have been able to use photographs and video footage to track a meteorite's path across the sky before it hit the ground. These records show that most meteorites come from the asteroid belt, which makes sense given the rocky nature of these objects. Astronomers believe that space, and the asteroid belt in particular, contains billions of small chunks of rock called meteoroids, each orbiting the Sun like a miniature planet. Every year, a small fraction of these rocks collide with Earth to become meteorites.

There is a limit to how long a small boulder can travel through space before it collides with something else. Any meteoroids that formed at the same time as the solar system would have crashed into a planet or an asteroid long ago. The meteoroids that exist today must have formed quite recently, and another piece of evidence confirms this. As meteoroids travel through space, they are bombarded by cosmic rays—energetic particles that come from the Sun and elsewhere in our galaxy. Cosmic rays react with atomic nuclei in the surface layers of a meteoroid, producing characteristic isotopes that can be identified when the meteoroid lands on Earth. The longer a meteoroid spends in space, the more cosmic ray products build up in its surface layers. Measuring the amounts of these materials shows that a typical meteoroid spends only a few tens of millions of years in space before landing on Earth.

Cosmic rays can penetrate only the outer few meters (several feet) of a rocky body. This suggests that meteoroids were once part of larger asteroids, buried too deeply below the surface for cosmic rays to reach them. How do meteoroids escape from their parent asteroid into space? The most likely explanation is that meteoroids are the debris produced when asteroids collide. Asteroids move at speeds of several kilometers (several miles) *per second* relative to one another. When two asteroids meet, the result tends to be catastrophic. Recently, astronomers were lucky enough to see the aftermath of two such collisions. In January 2010, a Lincoln Near Earth Asteroid Research (LINEAR) survey telescope discovered a fuzzy object resembling a comet in the asteroid belt. Closer scrutiny revealed that this was actually a large cloud of dust and gravel, apparently produced in a collision between two previously unknown asteroids early in 2009. In November 2010, the 100-km-wide (70-mile-wide) asteroid Scheila was hit by an object about 30 meters (100 feet) across, ejecting more than 660,000 tons of debris into space.

Collisions between asteroids can generate a steady supply of meteoroids, but this is not the whole story. When pieces break off an asteroid, they tend to follow more or less the same orbit as their parent. These meteoroids will still lie in the asteroid belt. Something must change their orbits before they can travel to Earth.

The Kirkwood gaps provide part of the answer (Figure 5.1). These narrow, almost empty regions of the asteroid belt are named after Daniel Kirkwood, who drew attention to them in the 19th century. Kirkwood

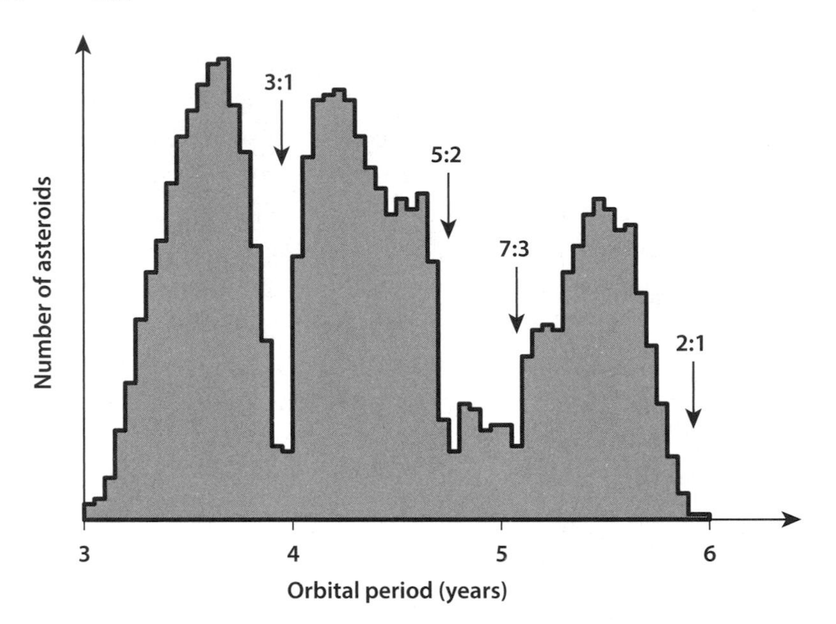

Figure 5.1. Kirkwood gaps. A simple graph showing the number of asteroids against orbital period reveals marked gaps (know as Kirkwood gaps) at periods that are in resonance with Jupiter's period. For example, an asteroid with a period of just under 4 years would complete 3 orbits for every 1 orbit Jupiter makes and would thus be in the 3:1 resonance.

gaps correspond to special orbital configurations called resonances. An asteroid located in the 3:1 resonance, for example, orbits the Sun three times every time Jupiter travels around the Sun once. As we saw in Chapter 3, gravitational perturbations from the planets cause the orbits of all bodies in the solar system to change slowly over time. Usually these changes consist of small oscillations that don't amount to much. However, an asteroid in a resonance behaves differently. Every time the asteroid passes Jupiter, it receives a small gravitational tug from the giant planet. Because of the resonance, the asteroid always meets up with Jupiter at the same point in its orbit, so the tugs always pull the asteroid in the same direction, slowly adding together over time. In the 1980s, planetary scientist Jack Wisdom used a computer to calculate these changes over several million years. He found that asteroids in a resonance soon develop highly elongated orbits that cross the paths of one or more planets. Meteoroids behave the same way. Any meteoroid that enters a resonance is likely to collide with the Sun or one of the planets within a few million years. Those meteoroids that hit Earth become meteorites.

One final piece of the puzzle remained—how do fragments from a collision between two asteroids end up in a resonance? This question was actually answered more than a hundred years ago by an amateur scientist called Ivan Yarkovsky, but nobody realized it at the time. Worse, Yarkovsky died shortly after publishing his findings in a pamphlet, and his discovery very nearly died with him. More than fifty years later, the noted astronomer Ernst Öpik recalled reading the pamphlet and remembered enough of the details to bring Yarkovsky's discovery to the world's attention.

What Yarkovsky discovered, and Öpik remembered, was that meteoroids don't just move due to the pull of gravity; their motion is also affected by sunlight. As a meteoroid travels through space, the side facing the Sun grows hotter than the side in shadow. Later, the meteoroid releases this heat energy back to space in the form of infrared radiation. Meteoroids are typically spinning, and they also take a while to warm up and cool down. By the time a meteoroid gives up its heat, it has rotated around and the Sun is no longer overhead (Figure 5.2). The same thing happens here on Earth, which is why it is usually hotter in the afternoon than in the morning even though the Sun is at the same elevation in the sky. On Earth, incoming sunlight and outgoing radiation have almost no effect beyond heating and cooling the planet. On a small meteoroid, however, the tiny push of sunlight and the recoil from infrared radiation are enough to continuously change the meteoroid's orbit around the Sun.

Millions of years after two asteroids collide, the fragments from the collision can drift far enough across the asteroid belt to reach a resonance where Jupiter's gravity forces some of them onto a collision course with Earth. Most of these fragments are small and land as meteorites that rarely cause any harm. However, a few fragments are much larger. Several thousand objects up to several kilometers (a few miles) in diameter have ended up on orbits that approach our planet. A collision with one of these near-Earth asteroids could have globally devastating consequences.

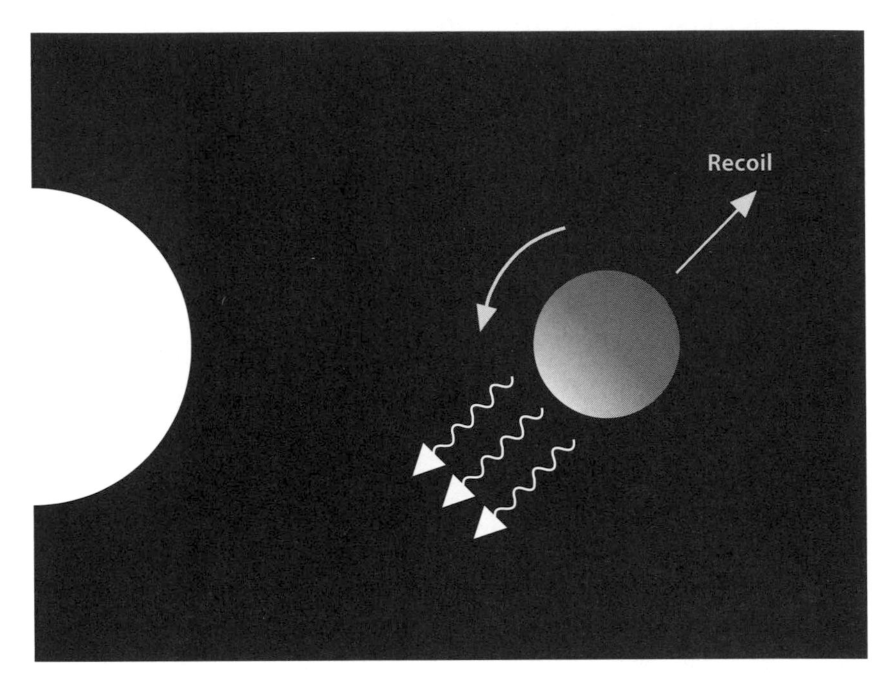

Figure 5.2. The Yarkovsky effect. The side of an asteroid that has been warmed by the Sun emits infrared radiation. This emission causes the asteroid to recoil very slightly in the opposite direction. Over time, this effect can noticeably alter an asteroid's orbit.

IRONS AND STONES

Meteorites come in two basic types. The easier to recognize are the shiny metallic specimens known as iron meteorites. These are chunks of nearly pure iron and nickel containing small amounts of gold, platinum, and other rare metals. When an iron meteorite is sliced open, cleaned with acid and polished, it displays beautiful interlocking patterns of shiny metallic crystals (Figure 5.3). Laboratory experiments show that crystals like these grow when molten metal cools very slowly and solidifies, forming minerals containing different amounts of iron and nickel. The compositions of the crystals suggest that they took millions of years to form. Iron meteorites must come from asteroids that were once so hot that the iron in them melted and settled to the center to form a metallic core.

Figure 5.3. An etched and polished section of the Gibeon iron meteorite, showing the structure of interlocking crystals, which is often called the Widmanstätten pattern. This sample is 4 cm (1.6 inches) wide. (J. Mitton)

Iron meteorites are actually not very common. They make up only a few percent of all the meteorites that fall to Earth. Most meteorites are stony, made of silicates and other rocky materials, with little or no metal. A typical stony meteorite contains many tiny round beads of rock, slightly smaller than the porous plastic balls that make up expanded polystyrene, or "Styrofoam" (Figure 5.4). These rocky beads are known as chondrules and are so characteristic that they have given their name to an entire class of meteorites called "chondrites." (Confusingly, a handful of chondrites contain no chondrules, but they have so much else in common with chondrites that scientists include them in the same class anyway.)

Chondrules are cemented together by a material called "matrix," made of microscopic grains of dust that seem to have a very different origin from the chondrules that lie right next to them. The matrix also contains a sprinkling of grains of diamond, silicon carbide, oxides, and silicates that are so different from anything in the solar system that they most likely came from somewhere else entirely. These "presolar grains" probably formed in the cool outer atmospheres of giant stars or supernova explosions, and then journeyed across our galaxy, entering the

Figure 5.4. A microscope image of loose chondrules. The scale marks are 1 mm apart. (Vatican Observatory, courtesy Guy Consolmagno)

solar system just when it was forming. Along with chondrules and matrix, chondrites also contain white, irregularly shaped particles called calcium-aluminum-rich inclusions (CAIs) (Figure 5.5). CAIs are about the same size as chondrules, but they are made almost entirely of exotic, ceramic-like materials with very high melting temperatures.

Chondrites come in several varieties. Unsurprisingly, the most common are called ordinary chondrites, and these account for four-fifths of all meteorites that fall to Earth. Ordinary chondrites are mostly made of the same rocky minerals that are found on Earth, especially silicates. Members of a second group are dubbed carbonaceous chondrites because they often contain significant amounts of carbon. Ordinary chondrites are typically dry, but the carbonaceous variety sometimes contain large amounts of water locked up inside clay minerals. Some carbonaceous chondrites even contain organic material, including amino acids and other molecules that form the basis of living organisms. A third group, the enstatite chondrites, are perhaps the most exotic of all. These meteorites formed in an environment where oxygen was extremely scarce. As a result, they contain bizarre materials such as nitrides and native silicon that never occur naturally in oxygen-rich places like Earth.

Some stony meteorites contain few chondrules but their basic chemical composition is very similar to that of chondrites. These achondrite

Figure 5.5. A microscope image of a section through the carbonaceous chondrite NWA 989, found in northwestern Africa in 2001. It shows both chondrules (round) and calcium-aluminum-rich inclusions (white, irregular). (Vatican Observatory, courtesy Guy Consolmagno)

meteorites look as if they come from asteroids that were heated to high temperatures at some point, causing their chondrules to melt and flow together. Some achondrites appear to come from the outer layers of asteroids that melted completely and formed a metallic core overlain by a shell of rocks similar to those we see in Earth's crust and mantle.

IDENTIFYING THE PARENTS

Although we are sure that most meteorites come from asteroids, it is surprisingly difficult to say which meteorite came from which asteroid. Very few asteroids have been seen at close range by a spacecraft, and we have only one sample of material collected from an asteroid—a small group of dust particles picked up from asteroid Itokawa by the Japanese Hayabusa mission.

In October 2008, astronomers were lucky enough to observe a tiny asteroid, named 2008 TC3, about 20 hours before it hit Earth. The 4-meter (13-foot) diameter object appeared as a brilliant fireball over Sudan before disintegrating in the Nubian desert, where scientists later collected 280 meteorite fragments. These meteorites tell us that the parent of 2008 TC3 was a highly unusual jumble of rocky chunks from many different bodies that were later assembled into a single asteroid. To date, 2008 TC3 is the only known asteroid that later fell as a meteorite.

One way to work out where other meteorites come from is to measure their spectrum in the laboratory and compare it with the spectrum of light reflected by different asteroids viewed through a telescope. In one case, this connection is easy to make. The HED meteorites—short for howardite, eucrite, and diogenite—are a group of about 1,000 objects with a uniquely characteristic spectrum. This spectrum is almost identical to that of Vesta, the largest body in the inner part of the asteroid belt. Astronomers have also found a nearby cluster of small asteroids with similar spectra that appear to be fragments from an ancient impact on Vesta. Some of these fragments lie close to a resonance, and these objects are probably the immediate parents of the HED meteorites, while Vesta itself is the grandparent.

Despite this success, it has been surprisingly difficult to find other matches. Many meteorites have a spectrum that doesn't correspond to any known asteroid, and most asteroid spectra don't match known meteorites. Fortunately, we now understand the reason why. Space is a very harsh environment. Asteroids and other objects without an atmosphere are constantly bombarded by cosmic rays, particles in the solar wind, and tiny meteoroids. This continuous assault transforms the outermost layer of an asteroid, breaking chemical bonds, and freeing metal atoms, which then coat the surface of the surrounding rock. The dust grains returned from asteroid Itokawa contained numerous microscopic particles of metallic iron and iron sulfide that changed the grains' outward appearance. Astronomers call this effect space weathering.

Space weathering makes an asteroid appear redder in color and washes out features in its spectrum. Meteorites are chunks that have broken off asteroids relatively recently, and their newly exposed surfaces have experienced space weathering for only a short time. This makes meteorites look fresh compared to asteroids. In a similar way, rocks on Earth change their appearance as they become old and weathered, although the weathering process is different. When geologists want to identify rocks, they usually look for a surface that has been exposed recently, or break open the rock to look at the pristine interior. Breaking open an asteroid is not a practical way of figuring out what it is made of. Instead, scientists take the opposite approach and artificially weather meteorites in the laboratory. When this is done, the match between meteorites and asteroids becomes much better and it is possible to estimate which type of asteroid a particular meteorite comes from.

Ordinary chondrites come from a class of bright asteroids called S-types that make up much of the inner part of the asteroid belt. Itokawa and several other asteroids visited by spacecraft belong to this class, and their prevalence in the inner asteroid belt—the region closest to Earth—may explain why so many meteorites are ordinary chondrites.

Carbonaceous chondrites appear to come from dark C-type asteroids in the middle and outer regions of the asteroid belt, many of which contain hydrated, clay-like minerals judging from their spectra. Curiously, these dark asteroids seem to be very common and yet carbonaceous meteorites are quite rare. These meteorites tend to be fragile, however, which suggests that many carbonaceous meteoroids break apart when they enter Earth's atmosphere and the pieces never make it to the ground. Micrometeorites, the smallest meteorites of all, waft gently down through the atmosphere, so even fragile objects survive. The great majority of micrometeorites look like carbonaceous chondrites rather than the ordinary kind.

Iron meteorites are associated with a relatively rare class of asteroids that have rather bland spectra, called M-types. However, some M-types appear to have water-bearing minerals on their surface. The existence of these minerals is hard to reconcile with the fact that iron meteorite parent bodies once grew hot enough to melt, and it seems likely that M-type asteroids represent more than one type of body and more than one group of meteorites.

Table 5.1. The basic classification of meteorites		
Chondrites	Carbonaceous	
	Ordinary	
	Enstatite	
Non-chondrites	Achondrites	Mars
		Moon
		Other
	Irons	

LUNAR AND MARTIAN METEORITES

Almost all meteorites originated in the asteroid belt, but some precious specimens came from somewhere closer to home. About 150 stony meteorites have been found that have an identical composition to the rocks brought back from the Moon by the Apollo astronauts. These small chunks of Moon rock were blasted off the lunar surface when the Moon collided with an asteroid or comet in the recent past. Even more surprisingly, we also have about 100 meteorites that seem to have come from Mars. Most of these rocks formed in the past two billion years, long after even the largest asteroids had cooled and become geologically dead worlds. These meteorites must have come from a geologically active planet instead, which makes Mars a likely candidate. Even more tellingly, the meteorites contain small amounts of gas trapped when the rocks formed, and the mixture of gases is identical to that in Mars's atmosphere measured by the Viking spacecraft.

Martian meteorites are particularly valuable because they provide us with samples from another planet without the tremendous expense of sending a spacecraft to fetch them. Because the rocks in these meteorites have a range of ages, they tell us about conditions on Mars over billions of years of its history. The minerals in one Martian meteorite are almost as old as the planet itself, and formed when Mars was much more like Earth than it is today. The lunar meteorites are also important. Although the Apollo astronauts returned some 382 kg (842 pounds) of

Moon rock, these all come from a handful of sites on the near side of the Moon. Lunar meteorites can come from anywhere on the lunar surface, giving us a more comprehensive picture of the Moon's history and composition.

A RARE AND PRECIOUS RESOURCE

Meteorites are valuable scientifically and greatly sought after. They are also uncommon, which makes hunting for them an arduous business. Most meteorites were never seen to fall, and superficially they are not readily distinguishable from other rocks on Earth. However, there is the one place in the world where identifying meteorites is easy even if collecting them is not: Antarctica, a continent almost entirely covered by ice several kilometers (a few miles) thick. Apart from the coasts and a few places where tall mountain peaks poke through the ice, the only rocks seen on the ground are those that fell from the sky.

Antarctica is a huge place, and meteorites are generally few and far between. Luckily, there are a few special locations where nature lends a helping hand by concentrating meteorites. Over thousands of years, the ice in Antarctica flows like a giant river of toothpaste, taking any meteorites with it. In some spots, the ice slows down to pass over an obstacle such as a mountain range. Over time, the wind erodes away the surface layers of ice as it flows upward, while meteorites are left behind. Eventually, meteorites become so numerous that dozens can be found in the space of a few kilometers (a few miles).

Every year for the past three decades, a small group of scientists has traveled to Antarctica's harsh interior, braving icy weather, isolation, and primitive living conditions to search for meteorites. Traveling by snowmobile, the meteorite hunters crisscross the ice, flagging each meteorite so it can be photographed and then packed away for transportation to a laboratory in warmer climes. In a good season, the hunters can find a thousand or more meteorites. To date almost 40,000 Antarctic meteorites have been discovered, more than all the meteorites found in the rest of the world put together.

Figure 5.6. The lunar meteorite ALHA81005, found in the Allan Hills area of Antarctica in 1981. It was the first lunar meteorite identified. It is almost identical to rocks collected on the Moon by Apollo astronauts and is a piece of ancient lunar crust. The cube at the lower left has sides 1 cm (0.4 inch) square. (NASA/JSC)

WHAT METEORITES CAN TELL US

Why go to this great effort, year after year? Isn't 40,000 meteorites enough? It is true that most newly discovered meteorites resemble others that have already been found—"just another ordinary chondrite" in many cases. Yet, such is the variety of meteorites that new kinds are being discovered every year. More importantly, meteorites contain a wealth of information about conditions in the early solar system, and each new meteorite discovery adds another piece to the puzzle.

Most rocks on Earth have been heated, melted, weathered, eroded, and recycled many times. Few rocks from the early days of Earth's history have managed to survive until the present day, and even these have been heavily modified over the eons. Worse still, Earth was probably so hot when it was young that it melted completely, destroying any material from the time before the planet formed. In many ways, searching for traces of ancient material on Earth is rather like trying to study archaeological ruins in the middle of a modern city where the ground has been dug up repeatedly to lay building foundations, sewer systems, and tunnels for an underground subway train.

In contrast, meteorites often contain a record of events early in the history of the solar system. Chondrites are particularly useful because they come from asteroids that never melted completely. They appear to be preserved grab bags of whatever material was floating through space at the time their parent bodies formed, rather like sedimentary rocks on Earth that formed when a random assortment of sand, gravel, and clay settled to the bottom of the ocean.

The large number of chondrules in chondrites tells us that these spherules were present in huge numbers early in the solar system. They are round because they were once droplets of hot rock and clearly show signs that they were heated strongly at some point in the past until they almost melted. Neighboring chondrules in the same meteorite often have very different physical and chemical properties, which means they could not have been heated while inside an asteroid. If they had been, they would have exchanged material or flowed together to form a homogeneous mass. Presumably, the heating events that produced chondrules occurred in space, and they must have been a common feature of the environment when the planets were forming.

The early solar system was clearly also a dynamic place, constantly mixing and exchanging material from one place to another. How else could matrix dust grains, full of water and fragile organic materials, end up jumbled together in the same meteorite with CAIs and chondrules, both of which were created at high temperatures, but under different circumstances?

Meteorites are not entirely pristine samples from the dawn of the solar system. The rocks in many chondritic meteorites have been partially modified when their parent asteroids were heated to hundreds of degrees at some point in the past. This heating was not enough to melt the asteroids, but it altered the rocks in their interior. In some cases, different meteorites from the same parent asteroid were heated by different amounts. This suggests that the rocks lay at different depths in a large asteroid that was hotter in the middle than near the surface. Later on, the asteroid must have broken apart in a collision, releasing rocks from various depths into space.

The parents of some meteorites, especially some carbonaceous chondrites, have been modified by liquid water. In some cases, the original minerals have largely been destroyed, replaced by clays and other hydrated minerals. It seems that these bodies once contained substantial amounts of ice that melted and reacted with dry rock in the asteroid's interior, forming new materials.

The HED meteorites tell us that Vesta experienced many of the same geological processes as Earth, including the formation of an iron core and volcanic eruptions on its surface. In some ways Vesta is more like a miniature planet than an asteroid. Dozens of other asteroids must have melted as well in order to produce the variety of iron meteorites we see today. Collisions later broke these asteroids apart, exposing their iron cores. Curiously most of the corresponding rocky mantle material from these bodies is missing, a puzzle we will return to in Chapter 13.

Perhaps meteorites' greatest contribution is their role as radiometric clocks, giving us a timeline for events during the formation of the solar system. The young solar system contained many radioactive isotopes that found their way into chondrules and asteroids, and ultimately meteorites. As we saw in Chapter 4, the amount and distribution of these isotopes can tell us when different components in the meteorites formed. In some cases, the radiometric clocks are incredibly precise, with an accuracy of one part in 10,000. This is equivalent to recalling the exact day on which an event took place nearly 30 years ago.

According to the radiometric clocks, the CAIs captured in meteorites are the oldest known objects that formed in the solar system. These tiny particles are 4.57 billion years old. Because they are so old, scientists often use CAIs as a guide to the age of the solar system itself, although it is possible that even older objects might be found in the future. Many CAIs formed within the space of only 100,000 years—a blink of an eye in the history of the solar system. Chondrules formed 1 to 3 million years after CAIs, and they have a wider range of ages. The fact that both types of particle are found in the same meteorite means CAIs spent several million years hanging around in space waiting for chondrules to appear, before they both found their way into the same asteroid.

Iron meteorites tell us about the history of asteroids that melted. The radiometric clocks show that many of these asteroids formed in the million years following the appearance of CAIs but before most chondrules formed. Some short-lived radioactive isotopes in the early solar system, especially aluminum-26, gave off large amounts of heat as they decayed. This heat would have been enough to melt any asteroids that formed in the first 1 to 2 million years of the solar system, which explains how the parent bodies of the iron meteorites came to be.

Asteroids that formed later, when most of the aluminum-26 had decayed away, would have experienced less heating—enough to thermally alter some rocks perhaps, and melt ice, but not enough to melt the whole asteroid. These asteroids became the parent bodies of chondrites, which are generally a little younger than the asteroids that melted.

Meteorites contain one more precious nugget of information: they tell us the mixture of chemical elements that was present in the solar system when the planets were being built. In the next chapter, we will examine this mix more closely, and see how the elements were created before the solar system was born.

SIX

COSMIC CHEMISTRY

Humans owe their existence to the rich variety of chemical elements that exist in the universe. Our solar system contains hydrogen to power the Sun, iron and silicon to build rocky planets, and carbon, nitrogen, and oxygen to form the building blocks of life. Almost 100 elements occur naturally in the solar system in varying amounts. Some, like hydrogen, oxygen, and iron, are abundant everywhere. Others, like gold, silver, and uranium, are much less common. The mixture of elements has remained almost constant since the solar system formed, apart from changes deep in the Sun's interior. In this chapter, we look at how the composition of the solar system was shaped by events elsewhere in the universe dating back to the Big Bang itself.

ELEMENT 43: FIRST A PUZZLE THEN A CLUE

In 1869, the Russian chemist Dimitri Mendeleyev published his famous periodic table, giving a sense of order to the bewildering array of elements known at the time. Mendeleyev arranged the elements by increasing atomic weight in such a way that elements in each row of his table had similar chemical properties. (Modern periodic tables are laid out somewhat differently, with similar elements occupying the same column instead of the same row.)

The concept of atomic weight that Mendeleyev used is different from atomic mass. Most elements have two or more stable isotopes, and a typical laboratory sample of an element contains a mixture of isotopes with different atomic masses. Take silicon as an example. It has three naturally occurring isotopes with atomic masses of 28, 29, and 30. Their relative proportions are about 92 percent, 5 percent, and 3 percent, respectively. The two heavier isotopes make the average *weight* per atom in natural silicon 28.09. Chemists could measure the atomic weights of elements long before isotopes were discovered.

Mendeleyev numbered the elements in the order they appeared in his table, starting with 1 for hydrogen, although it would be another half century before scientists realized the true significance of these atomic numbers. Mendeleyev wasn't the first scientist to draw up a table of this kind, but he paid more attention than his predecessors to the link between the order of the elements and their chemical properties. In a few cases, Mendeleyev decided to swap pairs of elements, such as nickel and cobalt, in order to get them in the right order for their chemical properties, even though the elements were no longer strictly ordered by atomic weight. As others had done, he also left gaps for elements that had yet to be discovered. Since new elements were being found at a rapid rate in the late 1800s, this foresight seemed reasonable and it also kept similar elements grouped together. Using his table, Mendeleyev was able to predict the properties of the missing elements by comparing the elements immediately around each gap.

Over the next few decades, the gaps in the periodic table were filled one by one until the table was almost complete. However, the gap between elements 42 and 44 (molybdenum and ruthenium) remained stubbornly vacant despite intensive searches by chemists around the world. If this missing element could not be found, the whole basis for the periodic table and understanding the elements would be undermined.

The situation was rescued in 1913 by a young British chemist, Henry Moseley. Moseley found that when an element is bombarded with electrons, it gives off X-rays with properties that are directly related to the element's atomic number. At about the same time, scientists realized that most elements exist as more than one isotope, each with the same atomic number and chemically identical, but containing atoms with different masses. These discoveries showed that atomic number has a fundamental physical meaning. It is atomic number rather than atomic weight that determines an element's chemical nature. Mendeleyev's periodic table was correct, even in the places where he had swapped pairs of elements around. The gaps in the table must also be real, and the hunt for element number 43 was renewed in earnest. Sadly, Moseley's promising career was cut short when he was only 27. He enlisted in the army after the outbreak of World War I and was killed in action in 1915.

Element 43 was finally discovered in 1937, in Italy, by Carlo Perrier and Emilio Segrè when they examined parts taken from a particle accelerator in California. It soon became apparent why the discovery had taken so long: the new element was radioactive, making it exceedingly rare in nature. In fact, the sample of element 43 found by Perrier and Segrè was not natural at all—it had been produced artificially when natural atoms of molybdenum were bombarded with subatomic particles in the particle accelerator. As a result, element 43 was named "technetium" from the Greek word for artificial. Scientists had known for several decades that atoms could change from one element to another when the nuclei of heavy elements like uranium broke apart during radioactive decay, but Perrier and Segrè's discovery showed that new elements could also be built from stable atoms.

Fifteen years later, in 1952, astronomer Paul Merrill found unmistakable signs that technetium was present in the atmosphere of a star. Technetium's most long-lived isotope decays away in only a few million years, much less than the age of the star, so any technetium that was present when the star formed would have disappeared long ago. Merrill's discovery proved that stars, just like particle accelerators, could make new chemical elements.

AN ABUNDANCE OF ELEMENTS

In Chapter 1, we saw how astronomers can use the Sun's spectrum to work out its composition. Most stable elements have been detected in the Sun, and their relative proportions have been measured to better than 10 percent in many cases. The remaining elements are probably present as well but in amounts too small to see. Here on Earth, scientists have measured the composition of meteorites with even greater precision.

The Sun has a remarkably similar composition to chondrites—the most primitive meteorites, which come from asteroids that have changed little since the formation of the solar system. The main exceptions to this rule are volatile gases such as hydrogen and helium that can escape easily from asteroids. The compositions of Earth, the Moon, and Mars are a little less similar to the Sun, but the broad trends are the same in each case. The most obvious way to explain this coincidence is that the Sun, the planets, and the asteroids were all made out of the same reservoir of matter when the solar system was forming.

Combining what we know about the Sun and meteorites provides a good guide to the composition of the solar system, and the material from which it formed. The first scientists to make a comprehensive estimate of the solar system's composition were Hans Suess, an Austrian geochemist, and the American chemist Harold Urey, who won the 1934 Nobel Prize for his work on isotopes. Their figures, published in 1956, combined actual measurements of many elements with educated guesses for others based on nuclear theory. Since then, scientists have measured more elements and improved estimates for others, but the overall pattern hasn't changed. Figure 6.1 shows the relative abundance of different elements in the Sun against their atomic number. The range of abundances is so great that the scale is a logarithmic one, labeled in factors of 10.

As the figure shows, heavy elements are generally rarer than light ones. A number of broad peaks and troughs are superimposed on this trend, together with a zigzag pattern. Working from left to right, there is a huge drop between helium (atomic number 2) and the next three elements, lithium, beryllium, and boron. Then there are two broad peaks. The first corresponds to what are called the alpha elements. These elements, such as carbon, oxygen, and neon, have nuclei that can be built entirely from alpha particles—helium nuclei consisting of two protons and two neutrons. The second broad peak centers on iron (atomic number 26) and includes metals such as chromium, nickel, copper, and zinc. As we will see, all of these features make sense when we understand how elements are manufactured in the universe.

We begin with hydrogen and helium, which are by far the most abundant elements in the Sun and most other stars. The extremely tenuous

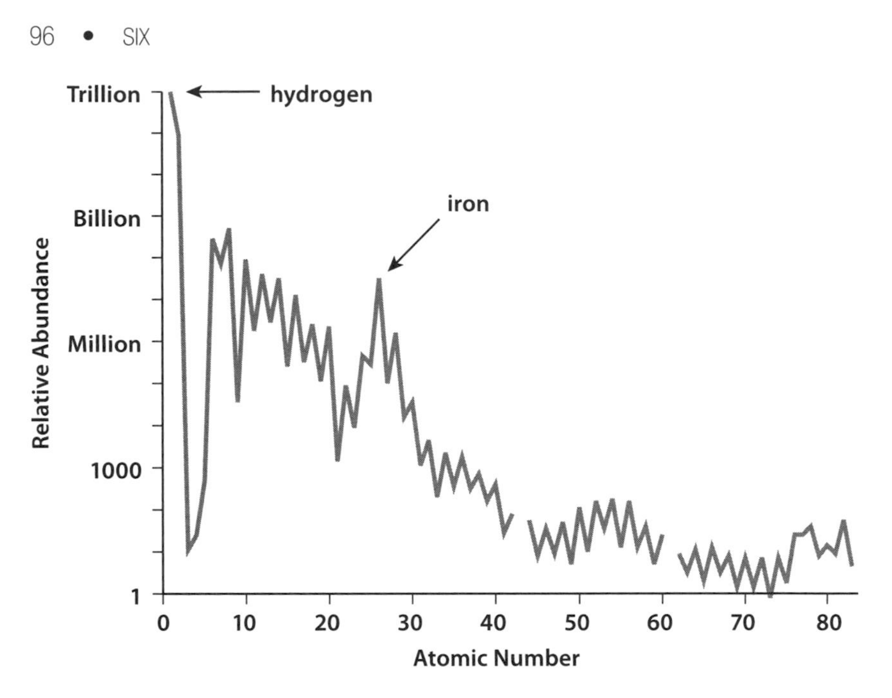

Figure 6.1. The relative abundance of the chemical elements in the solar system. The vertical scale is logarithmic. (Based on data from N. Grevesse, M. Asplund, and A. J. Sauval, *Space Science Reviews* 130 [2007]: 105–14)

gas that exists in the space between stars is also mostly made of hydrogen and helium. To understand why the two lightest elements are so abundant, we need to journey back in time to the earliest moments of the universe immediately after the Big Bang.

THE FIRST ELEMENTS

In its infancy, the universe was very different than it is today. There were no galaxies or stars, no planets or asteroids, nor even any atoms. Instead, shortly after the Big Bang, the universe was a roiling cloud of subatomic particles, positively charged protons, negatively charged electrons, and neutral neutrons. Particles traveled at tremendous speeds, bumping into other particles over and over again. Particles of electromagnetic radiation called photons blinked in and out of existence as they were generated by some subatomic particles and quickly absorbed by others.

The universe was also changing, expanding and cooling rapidly. In the first split second after the Big Bang, interactions between even more primitive particles established the ratio of protons to neutrons and electrons. Neutrons, being heavier than protons and less stable, formed less often, so that roughly one neutron was created for every seven protons. The oppositely charged electrons and protons formed in equal numbers so the universe became electrically neutral overall.

When the temperature had fallen a little, colliding protons and neutrons began to fuse together, attracted by strong nuclear forces. These aggregates and leftover protons would ultimately become the nuclei of atoms, so we call these nuclei from now on. Electrons, being much lighter than protons and neutrons, traveled at higher speeds—too fast to interact significantly with their heavier cousins at this stage.

The first nuclei to form fell apart again almost immediately when they were hit by an energetic photon. Nuclei came and went in rapid succession, always returning to the cloud of protons and neutrons swirling around them. After about three minutes, the universe had cooled to the point at which photons were no longer powerful enough to break nuclei apart and the nuclei began to grow larger. At first, protons and neutrons combined to form deuterons, two-particle nuclei consisting of a proton and neutron. These absorbed more protons and neutrons to produce larger nuclei. Nuclei are all positively charged since they contain protons, so they strongly repel other nuclei. However, collisions were energetic enough to overcome this repulsive force, allowing nuclei to get close enough for short-range nuclear forces to bind the protons and neutrons together.

Some nuclei are more robust than others. Nuclei containing an even number of protons tend to be more stable than those with an odd number, which accounts for the zigzag pattern in Figure 6.1. Greater stability means that a nucleus can form more easily in the first place and put up a stronger resistance when threatened with destruction in a collision. The combination of two protons and two neutrons in a helium-4 nucleus (also known as an alpha particle) is particularly stable. Almost all the neutrons quickly found their way into alpha particles, along with a corresponding number of protons, leaving a large number of leftover single protons. Adding another proton to an alpha particle yields a highly unstable nucleus, and the same is true when two alpha particles merge together. As a result, the growth of nuclei virtually stalled at this point, except for a few lucky exceptions that managed to cobble together six or seven protons and neutrons.

After only a few minutes, the universe cooled so much that the positively charged nuclei were no longer able to overcome the repulsive forces between them and fuse together. At this point, all nuclear reactions ceased. Roughly three-quarters of the mass remained in lone protons, with most of the rest in alpha particles. A small number of deuterons and helium-3 nuclei—containing only one neutron—were also present. Only trace amounts of material existed in nuclei larger than alpha particles.

For the next few hundred thousand years, electrons stood aloof from the proceedings, still traveling too fast to interact noticeably with nuclei. After about 400,000 years, the universe had cooled to a few thousand degrees. At these temperatures, positively charged nuclei were finally able to team up with negatively charged electrons in arrangements that were electrically neutral. These were the first atoms. Atoms with one proton became hydrogen, including the deuterons, which became deuterium or "heavy hydrogen." Atoms with two protons, including the alpha particles, became helium. Those rare atoms with three protons became lithium.

The young universe must have been an exotic place in many ways, but chemically it was extremely dull. Only three elements existed: hydrogen, helium, and lithium. Helium refuses to bond with any element, including itself, while hydrogen and lithium form a very limited set of chemical compounds. The early universe contained no iron or silicon to build planets, no carbon or nitrogen to form living organisms, no oxygen for animals to breathe. The formation of all these elements had to await the birth of stars.

COOKING IN THE STELLAR FURNACE

Stellar interiors are one of the few places that can mimic the hot, dense conditions in the first few minutes after the Big Bang. Instead of lasting for only a few minutes, the nuclear furnaces in stars can burn for billions of years. This makes them ideal places to synthesize new elements out of existing ones. New elements form quite slowly inside a star, but the universe contains trillions of stars, and stars have been steadily creating new elements for about 13 billion years. More elements are created when massive stars explode. Over time, stars have converted roughly 2 percent of all the hydrogen and helium in the universe into heavier elements.

How exactly do new elements form inside stars? Nuclear physicists and astrophysicists began to tackle this question in earnest in the 1950s. One of these was Fred Hoyle, who would become well known for his strong but ultimately futile opposition to the Big Bang theory of the universe. Hoyle began his research career in the 1930s as a graduate student at Cambridge and made several important discoveries before he turned his attention to how chemical elements form inside stars. In 1953, the American nuclear physicist Willy Fowler introduced Hoyle to Margaret and Geoffrey Burbidge, who were trying to make sense of a star with a strong magnetic field and a strange spectrum as well as a very unusual composition. The Burbidges wondered whether particles accelerated by the star's powerful magnetism might be creating new elements. Fowler was intrigued by the idea and proposed that the four should team up to investigate how elements are produced inside stars.

When Suess and Urey published their table of abundances in 1956, the quartet intensified their efforts. A year later, the Burbidges, Fowler, and Hoyle (universally known to astronomers as B^2FH) published a comprehensive account of nuclear processes in stars and how they have produced the mix of elements we see in the solar system.

Nuclear reactions can take place in the centers of stars because the pressure and temperature are extremely high. Electrons are stripped away from atoms, leaving a plasma composed of naked nuclei and free electrons, just like matter in the first few moments after the Big Bang. Particles repeatedly run into one another at tremendous speeds. Most of the time, the nuclei's positive charges hold them apart so they stop short of actually fusing together. Every once in a while though, two nuclei get close enough for nuclear forces to overcome this repulsion, causing the particles to fuse together into a larger nucleus. Nuclear fusion generates highly energetic gamma radiation. The gamma rays percolate outward, being absorbed and reemitted many times, and gradually losing energy in the process. After thousands of years, the radiation reaches the star's surface and escapes as visible light. The star shines.

Figure 6.2. *From left to right*: Margaret Burbidge, Geoffrey Burbidge, Willy Fowler, and Fred Hoyle. This photograph shows them with the model steam train presented to Fowler at the conference organized in July 1971 in honor of his 60th birthday, 14 years after the publication in *Reviews of Modern Physics* of their famous paper on the creation of chemical elements in stars. (Courtesy Donald Clayton)

Most stars generate energy by converting hydrogen into helium. This can happen in one of two ways. In relatively small stars like the Sun, helium is built piece by piece, one hydrogen nucleus at a time in a process called the proton-proton chain (Figure 6.3). Initially, two hydrogen nuclei (protons) fuse to form a deuteron, with one proton changing into a neutron in the process. A third proton is added, and then a fourth, to form a helium nucleus. The temperature is too low for larger nuclei to form and the process stops there—at least for the time being.

Stars that are more massive than the Sun can make helium in another way as well. The process starts with the commonest isotope of carbon, carbon-12, which contains six protons and six neutrons. Four protons are added to a carbon-12 nucleus one at a time, ultimately generating an unstable nucleus containing 16 protons and neutrons. This spontaneously breaks apart into a carbon-12 nucleus and an alpha particle. The net result is that four hydrogen nuclei have fused together to form one helium nucleus, just like in the proton-proton chain. Meanwhile, the original carbon-12 nucleus is regenerated so it can continue to act as a

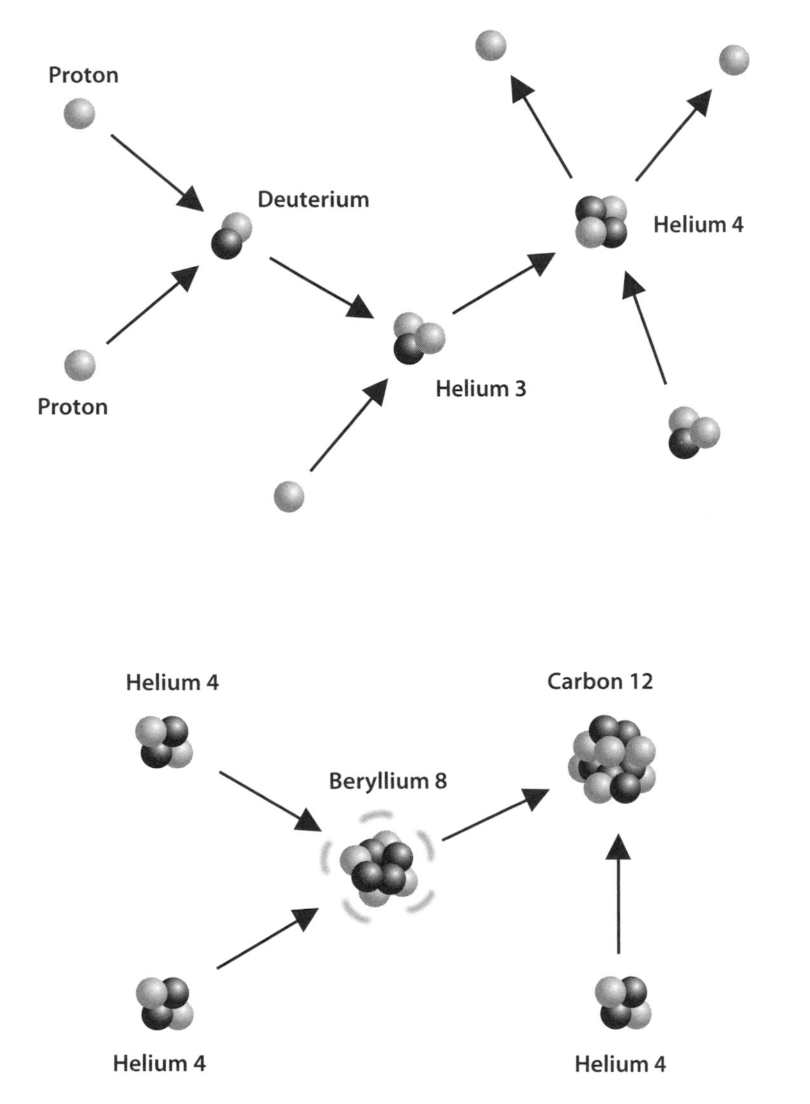

Figure 6.3. Simplified schematic diagrams of two of the main nuclear processes that take place in stars. *Top*: The production of helium-4 by the proton-proton chain process. *Bottom*: The production of carbon-12 by the triple alpha process.

catalyst, helping to keep the nuclear reactions going without being used up in the process. In the intermediate reactions, isotopes of nitrogen and oxygen are created so the whole process is called the CNO cycle after the chemical symbols for carbon, nitrogen, and oxygen.

You may wonder where the carbon-12 nuclei came from in the first place. The answer is that the carbon had formed earlier inside different stars and was subsequently ejected into space when those stars reached the ends of their lives. The first stars to form could not have used the CNO cycle to burn hydrogen because the universe did not contain any carbon at that point. Later generations of stars, which formed out of material enriched with carbon, were able to make use of the CNO cycle.

Throughout most of their lives, stars maintain a delicate balance between explosion and collapse. The energy released by nuclear reactions in the center of a star heats material to millions of degrees, increasing the internal pressure and causing the star to expand. Left unopposed, this tremendous pressure would blow the star apart. At the same time, the star's immense gravity seeks to crush its matter ever more tightly together. Nature manages to find a balance between these extremes. If a star begins to shrink, material in the center grows hotter, nuclear reactions speed up, more energy is released, and the star expands again. If the star expands too much, the interior cools, nuclear reactions slow down, and the star's gravity causes it to shrink.

Most stars keep this balancing act going for billions of years. Over time, hydrogen in the core of the star is gradually converted into helium, and the star grows a little denser and a little hotter as a result. However, the balance between pressure and gravity is always maintained. Tiny red dwarf stars last especially long. These stars are extremely frugal with their hydrogen fuel, shining feebly compared to the Sun, and they continually mix fresh hydrogen from their outer layers into the interior to replace material that has already burned. As a result, the smallest red dwarfs will shine for trillions of years, much longer than the age of the universe so far.

Larger stars shine more brightly and consume their hydrogen fuel rapidly. Inevitably, there comes a point when all the hydrogen in the core of a star is exhausted. As the nuclear fires grow dim, the temperature in the stellar interior falls. The core is less able to hold itself up against the downward pressure of the overlying layers, and gravity makes the core fall in on itself. The release of gravitational energy raises the temperature around the core and nuclear reactions start up again in the surrounding shell of hydrogen-rich gas. This new burst of energy heats the outer layers of the star, building up pressure that makes the star swell to many times its former size. The star becomes a red giant.

Eventually, the temperature and pressure in the core become so high that helium nuclei begin to fuse together. When two helium nuclei merge, they form a beryllium-8 nucleus. However, beryllium-8 nuclei are incredibly unstable, falling apart into two alpha particles again in only a few trillionths of a trillionth of a second. Almost every time a beryllium-8 nucleus forms it immediately breaks up again and nothing is gained. Once in a while, however, a third alpha particle hits a beryllium-8 nucleus during its fleeting existence. These nuclei fuse together, making a stable carbon-12 nucleus and releasing energy (Figure 6.3). As the stellar core continues to shrink due to gravity, nuclei collide with one another at a faster and faster rate, and soon the star is burning helium into carbon at a fast enough pace to halt further collapse.

Astronomers call helium burning the "triple alpha process" because three alpha particles are needed to make each carbon nucleus. Occasionally, a carbon nucleus absorbs an additional alpha particle to form an oxygen nucleus. Interestingly, although beryllium-8 has what seems like an impossibly short lifetime, the triple alpha process works effectively because of a fortuitous coincidence in the properties of helium, beryllium, and carbon nuclei—a phenomenon first identified by Fred Hoyle. Under marginally different circumstances, beryllium-8 would be much less likely to merge with an alpha particle, rendering the triple alpha process unviable. In such an alternate universe, carbon, oxygen, and all the other elements would never form, and life as we know it could not exist.

Most stars are too small to make heavier elements by fusing together carbon and oxygen nuclei. Once helium in the core is used up, nuclear reactions in the center of the star cease. The core shrinks and becomes denser until particles are pressed so tightly together that the core can shrink no more. Away from the core, layers of helium and hydrogen gas continue to burn. The energy released in these layers makes the star swell to an even greater size. These stars are so large that they swallow up any planets orbiting nearby. This is probably the ultimate fate of Mercury and Venus, and possibly Earth, when the Sun reaches this stage in its evolution in about 5 billion years from now.

As the supplies of hydrogen and helium in the burning layers wax and wane, the star's nuclear reactor produces energy in irregular pulses, like the sputtering engine of a car whose fuel tank is almost empty. These pulses gradually blow the outer layers of the star into space until all that is left is an inert husk of carbon and oxygen—a dead, ultra-dense white dwarf star that is destined to spend eternity gradually cooling and fading away.

BUILDING HEAVIER ELEMENTS

Nuclear reactions within red giant stars produce an intermittent stream of neutrons. Since neutrons are neutral, they are not repelled by other particles and they are readily absorbed by nuclei. Nuclei that take in an extra neutron often become unstable as a result. They then decay into a different element, typically with a higher atomic number, before the next neutron comes along. In this way, heavier and heavier elements are gradually built up. B^2FH called this the s-process, where the "s" stands for "slow." It is an important source of rare, heavy elements, as well as technetium.

Red giants typically undergo several upheavals in their later years as episodes of helium burning come and go. During these upheavals, carbon, oxygen, and heavy elements created by the s-process are dredged up from the stellar interior into the surface layers of the star. This is why Paul Merrill was able to detect technetium in such a star. Much of this chemically enriched material is blown off into space in the form of gas and tiny dust grains that will form the next generation of stars.

Stars much more massive than the Sun are not destined to become white dwarves. They have a more dramatic fate. As the central pressure and temperature rise in these stars, carbon and oxygen nuclei begin to fuse together, forming the heavier alpha elements—neon, magnesium, silicon, argon, and calcium—in a succession of reactions. Much of the energy generated by these reactions is carried away by neutrinos—ghostly particles that rarely interact with other matter. Most neutrinos quickly pass all the way through the star to the surface, escaping into space and stealing energy as they go. As a result, nuclear reactions in the star's core have to go faster and faster to provide enough energy to counteract the force of gravity. The nuclear burning stages get shorter as each new fuel ignites. Whereas hydrogen burning lasts for millions of years, the final stage, in which silicon is converted into iron and nickel, lasts only a few days.

The star has reached a moment of crisis. So far, every element the star has manufactured could serve as new fuel when the old fuel ran out. Fusing iron, nickel, and heavier elements consumes more energy than it produces. The star has run out of energy sources. Inevitably, gravity reasserts itself and the core of the star plummets inward. Protons and electrons in the core are squeezed ever more tightly together until they merge to form neutrons. The neutrons themselves put up resistance when tightly packed together and, if the core is not too massive, the pressure they exert brings the collapse to a sudden halt. The core becomes a neutron star, an object as massive as the Sun yet only about 20 km (12 miles) in diameter. In the most massive stars, even the neutrons are incapable of holding back gravity. The cores of these stars become black holes, ultra-dense objects whose gravity is so strong that not even light can escape.

SUPERNOVAE

As the core of a massive star collapses, the layers above it become compressed and heated by neutrinos. The dramatic rise in temperature and pressure sets off a furious burst of nuclear reactions. The star generates as much energy in a few moments as it did in its entire lifetime up until this point. This burst of energy literally blows the star apart, flinging the outer layers into space at a sizeable fraction of the speed of light. The star has become a supernova.

Most of the iron-rich material in the core remains trapped there during the collapse. However the sudden outburst of nuclear reactions in the layers above generates large amounts of heavy elements, especially iron, which are rapidly dispersed into space. These reactions also produce huge numbers of neutrons that are immediately absorbed by neighboring nuclei. The new nuclei are mostly unstable, but the flood of neutrons is such that they do not have time to decay before absorbing additional neutrons. Many exotic and extremely heavy nuclei are formed as a result, including uranium and plutonium. B^2FH termed this the r-process, where "r" stands for "rapid."

Supernova explosions can be triggered in another way as well. Many stars are part of binary systems—two stars orbiting about their common center of mass. The more massive member of the pair evolves faster. It is the first to exhaust its hydrogen fuel, become a red giant and ultimately a white dwarf. When its partner later puffs up to become a red giant, some gas from the giant's outer layers is pulled onto the white dwarf by the dwarf's gravity. If enough extra mass builds up on the white dwarf, runaway nuclear reactions begin, generating a sudden burst of energy that blows the white dwarf apart. These supernovae generate heavy elements by the r-process and spew these elements into the surrounding space, just like supernovae involving a single star.

Supernovae are rare. No supernova has been seen in our own galaxy for the past 400 years. In 1987, astronomers were treated to the next best thing when a supernova (SN1987A, Figure 6.4) appeared in one of our galaxy's nearest neighbors, the Large Magellanic Cloud. The progenitor, a single giant star, had been observed from time to time for a century beforehand and had shown no sign that anything unusual was about to happen. Then, on February 23, it brightened dramatically in the space of only a few hours. At the same time, detectors buried deep underground on Earth registered a burst of neutrinos passing through our planet, the first time neutrinos had been detected coming from an object in space other than the Sun.

Over the next three months, the supernova brightened steadily before fading away again. At its peak, it could be seen clearly with the naked eye in the southern hemisphere. The light came from an expanding cloud of gas that was all that remained of the outer layers of the star, together with energy released by the decay of radioactive cobalt and nickel that formed during the explosion. After several months, the gas cloud had expanded enough to reveal its inner regions. Spectra of the cloud clearly showed that it was full of heavy elements, dramatic proof that new elements are made inside stars and ejected into space when stars die.

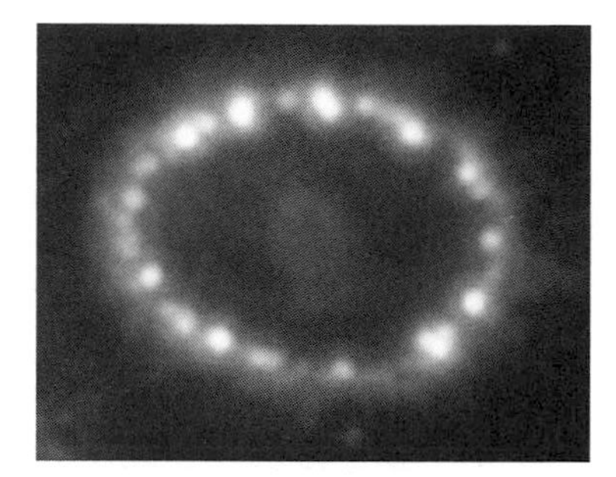

Figure 6.4. An image of the region around the supernova SN 1987A taken by the Hubble Space Telescope in 2003, 17 years after the explosion took place. The ring of bright blobs, which is about a light-year across, is where the shock wave created in the explosion has slammed into a ring of gas previously shed by the star, probably about 20,000 years earlier. The fainter nebula in the middle of the ring is an expanding cloud of glowing debris ejected by the blast. It is being heated by the decay of radioactive isotopes, principally titanium-44. (NASA, P. Challis, R. Kirshner [Harvard-Smithsonian Center for Astrophysics] and B. Sugerman [STScI])

The material expelled by dying stars is steadily enriching the gas that forms each new generation of stars. By the time the solar system formed, roughly 2 percent of the material in our region of the galaxy had already been converted into elements heavier than helium. In the next chapter, we will see how the tenuous mix of gas and dust in interstellar space evolved into our Sun.

SEVEN

A STAR IS BORN

> Recipe for a star: Take 10^4 solar masses of molecular gas. Sprinkle liberally with carbon and silicate dust spiced with metals. Freeze to 10 K and stir well until mixture is frothy. Hammer until lumpy. No need for oven; stars will form and bake themselves. Watch out for hot bubbles flying from pot.
>
> —*James Kaler, "Cosmic Clouds"*

A CHILD OF THE MILKY WAY

When we look up at the stars on a clear night, we see several hundred of the Sun's nearest and brightest neighbors forming the familiar patterns of the constellations. If the night is particularly dark, several thousand other stars are visible to the naked eye, as well as a hazy band of light stretching across the sky from horizon to horizon. This band, composed of billions of individual stars, is our Milky Way galaxy. Billions more stars lie concealed behind swathes of dust or are simply too dim and too far away to be seen. Our Sun is just an unremarkable star among all these billions.

The Milky Way appears to be fairly ordinary as galaxies go. While it isn't easy to work out exactly what our galaxy would look like from the outside, decades of effort by astronomers have given us a remarkably detailed picture of the Milky Way and its contents. We have every reason to believe that the Milky Way resembles countless other barred spiral galaxies seen throughout the universe.

The main component of the Milky Way is a rotating disk of stars some 100,000 light-years across but only about 1,000 light-years thick. From our vantage point on Earth, we see this disk edge-on, which is why the Milky Way appears as a narrow band across the sky. Between the stars lies an extremely tenuous mixture of gas and fine dust grains called the interstellar medium. The disk of stars is only about 1,000 light-years thick but becomes thicker near the Milky Way's center, where a bar-shaped bulge of densely packed stars surrounds a supermassive black hole at the heart of the galaxy. Enveloping the thin stellar disk is an extended disk of gas about 10 times thicker.

The Milky Way's most striking feature is a pattern of bright arms that wind outward in loose spirals through the disk heading away from the central bulge. These spiral arms are not permanent structures but temporary concentrations of stars, gas, and dust that sweep through the disk, like the waves that travel around a sports stadium when thousands of spectators stand up and sit down in unison. The increased density of gas and dust in the spiral arms provides ideal conditions for new stars to form. The spiral arms are speckled with the brilliant blue-white beacons of massive young stars, and hydrogen clouds made to glow pink by intense ultraviolet radiation from newly formed stars embedded inside them.

Today, stars are forming in the Milky Way at a rate equivalent to one solar-mass star every year. Judging by the age of its oldest members, the Milky Way has been giving birth to new stars for over 13 billion years, beginning less than a billion years after the Big Bang. The galactic halo—a roughly spherical region that surrounds the disk—contains many stars that are particularly ancient. One of these, a red giant star that goes by the catalog number HE0107-5340, has an age of 13.2 billion years, three times older than the Sun. Many halo stars belong to globular clusters—dense ball-shaped groups containing hundreds of thousands of stars whose origins remain somewhat mysterious. Unlike the galactic disk, these globular clusters are devoid of gas and dust, and new stars cannot have formed within them for a very long time.

WHERE STARS ARE BORN

The solar system lies at the inner edge of one of the Milky Way's spiral arms, a little more than halfway out from the center of the galaxy (Figure 7.1). Currently, we are passing through an unusual bubble, 300 light-years across, where the interstellar gas is 5 to 10 times less dense than average. Spiral arms sweep past the Sun roughly every 100 million years, taking about 10 million years to do so. At the same time, the Sun is traveling around the center of the galaxy once every 230 million years or so. Having completed 20 to 30 orbits so far, it has been in and out of spiral arms between 40 and 50 times. Clearly, the Sun's surroundings have changed a good deal over time, and it may be impossible to say

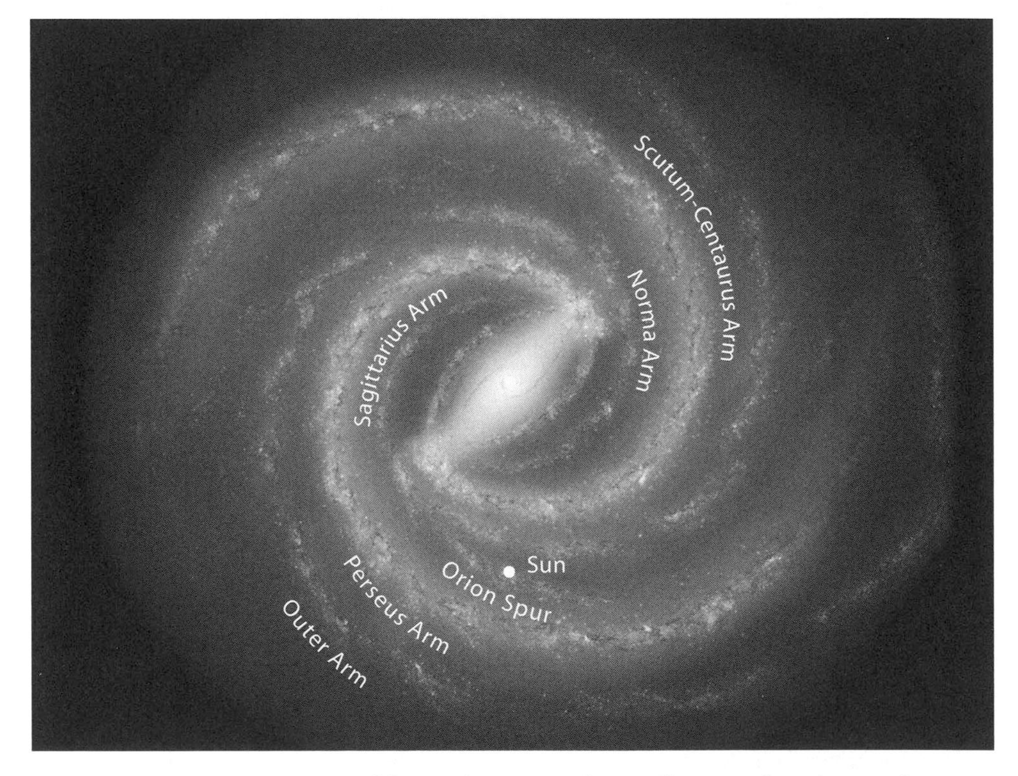

Figure 7.1. An artist's conception of the spiral structure of our Milky Way galaxy showing the current location of the Sun in a small, partial spiral arm, called the Orion Spur. The two major arms have the highest densities of both old and young stars. The less prominent arms consist mainly of gas and areas of star formation. (NASA/JPL-Caltech/R. Hurt [SSC-Caltech])

exactly where in the Milky Way our solar system formed. However, we can learn a lot about the Sun's birth environment, and how it shaped the solar system, by looking at places where stars are forming today.

One of the most intensely studied stellar nurseries lies about 1,500 light-years away in the constellation of Orion. Lurking behind the bright foreground stars that mark the familiar shape of Orion lie two giant molecular clouds, invisible in ordinary light but easily seen using infrared detectors. Molecular clouds are among the largest and most massive structures in the galaxy, each containing 10,000 to a million times the mass of the Sun. These clouds get their name because they are largely made of molecules of hydrogen rather than the individual hydrogen atoms that typically make up the interstellar medium. More than a hundred other molecules, including complex organic chemicals, have also been seen in molecular clouds.

Molecular clouds are the coldest and densest places in the interstellar medium. Shielded from outside light and radiation by layers of dust, the temperature inside a molecular cloud hovers only 10 degrees above absolute zero (−263°C or −441°F). If our Sun were inside a molecular cloud today, every other star in the sky would be invisible, hidden by the surrounding dust. The sheltered conditions inside a cloud allow atoms to combine in twos, threes, and larger groups, spicing the material that will become stars and planets with a rich variety of chemicals. The cold, dense environment also turns out to be a perfect incubator for infant stars.

Most of the star-forming activity in Orion is concealed behind its dark clouds. We can glean some information using telescopes that detect infrared radiation or microwave and millimeter radio waves, since this radiation can penetrate the dust where visible light cannot. However, nature has kindly provided us with a remarkable window into part of the star-forming region, allowing us to see the brightly glowing Orion nebula and newborn stars embedded within it. Here, a loose cluster of several thousand stars less than a million years old occupies a region less than 10 light-years across. To put this in context, if our middle-aged Sun is equivalent to a solitary 40-year-old adult, the cluster in Orion is like an overcrowded nursery of 3-day-old babies.

Blazing at the heart of the Orion cluster is a dense group of stars only 300,000 years old centered on four massive, white-hot stars called

Figure 7.2. A composite, near-infrared image of the Trapezium star cluster in the Orion Nebula. (ESO)

"the Trapezium." The brightest member of this quartet is 40 times more massive than the Sun and 200,000 times as luminous. Strong winds and powerful ultraviolet radiation produced by the Trapezium stars have cleared a cavity in the dust, leaving a glowing bubble of gas. This bubble has burst through the front of the molecular cloud, allowing us to peer inside. The scene revealed is a highly productive star factory, set against a background of luminous gas woven into an intricate pattern of filaments and streamers.

The molecular clouds in Orion contain thousands of young stars with masses ranging up to 40 solar masses, but not all molecular clouds produce stars that are so massive or so numerous. Four hundred light-years from us, in the constellations Taurus and Auriga, dark spots and filaments silhouetted against a starry background hint at the presence of a smaller stellar nursery. Closer scrutiny reveals all the telltale signs of star formation but no massive, extremely luminous stars like those in the Orion cluster. Most of the young stars in Taurus and Auriga exist in small groups containing only 20 to 30 objects rather than the thousands of newborn stars in Orion.

FIRST STEPS TO A SOLAR SYSTEM

A molecular cloud is made of turbulent, magnetized gas, laced with grains of dust. Occasionally, turbulent motions within the gas generate clumps and filaments of material. Once a clump forms, gravity can take hold. Portions of the clump begin to shrink and grow denser to form "cores"—gaseous seeds that will ultimately become stars. Cores may form in large swarms, small groups, or singly depending on the size, mass, and density of their parent cloud.

Cores normally remain hidden inside their dark clouds, but occasionally they become exposed to view. A good example is the Eagle Nebula, a star-forming region about 6,500 light-years away from Earth. In 1995, astronomer Jeff Hester and his colleagues used the Hubble Space Telescope to take a close-up picture of a small part of the Eagle Nebula. It would become one of the most iconic Hubble images ever produced, popularly known as the "pillars of creation" (Figure 7.3). The pillars in question are columns of dusty gas protruding from a molecular cloud, visible as dark silhouettes against a bright background of gas made luminous by radiation from nearby hot, young stars. Ultraviolet radiation from these stars is gradually eating deeper into the molecular cloud, breaking molecules apart and stripping electrons from their atoms, forming hot plasma that streams away into the surrounding space. This process, called photoevaporation, eventually destroys a molecular cloud, putting a halt to star formation. Photoevaporation can also affect nascent planetary systems, as we will see.

Dense portions of a molecular cloud hold out against photoevaporation longer than the more tenuous regions. In the Eagle Nebula, the thinner gas has already eroded away, leaving behind the pillars, together with numerous dense cores that appear as bumps on the pillars and as

Figure 7.3. A Hubble Space Telescope image of pillar-like clouds of dense gas and dust in the Eagle Nebula, which have been sculpted by photoevaporation in this star-formation region. The tallest pillar is about 4 light-years long. (NASA, ESA, STScI, J. Hester and P. Scowen [Arizona State University])

detached blobs. Appropriately, astronomers call these objects "EGGS," short for evaporating gaseous globules. Stars are already visible inside some of the EGGS. Others may never hatch as photoevaporation erodes away too much mass for a star to form. Cut off from the parent cloud, these EGGS can no longer pull in more gas, and they are destined to evaporate completely.

The Sun must have begun life as an extended ball of gas and dust similar to the cores we see in the Eagle Nebula. Turbulent motions within the Sun's parent molecular cloud gave rise to a core that was rotating slowly when it first formed. Gravity pulled material inward, compressing the gas and generating heat. Initially most of this heat escaped into space as infrared radiation. As the center of the core grew denser, some radiation became trapped and the temperature started to rise. As the core shrank, its rotation rate increased. Much of the inflowing material overshot the center and collided with material coming from the opposite direction, forming a flattened disk of gas and dust rotating around a dense central clump.

Eventually, gas in the center of the core reached a temperature of about one million degrees. At this point, the first nuclear reactions began. Temperatures were still too low for ordinary hydrogen nuclei to fuse together, but deuterium began to burn instead. Deuterium fusion released enough energy to halt further contraction, at least temporarily. Some 10,000 years after the core first began to collapse it had become a protostar, an important milestone on the way to becoming a true star. Deuterium is rare, however, making up only about 1 atom in every 100,000, so the supply was soon exhausted and the core began contracting once more.

Had any astronomers been around at this time, they would have unable to see the infant Sun at this stage. Only the outer parts of the dense envelope surrounding the Sun would have been visible, extending out for 1,000 to 10,000 astronomical units (AU). An observer may have seen streams of gas emerging from opposite sides of the dusty envelope. Strange as it may seem, a growing star has to expel about 10 percent of the material falling on to it in order to survive. If this didn't happen, a star would end up spinning faster and faster as it contracted until it was torn apart. Instead, a fraction of the material flowing in toward a young star is redirected by its magnetic field and pumped outward into high-speed jets, taking much of the star's rotational energy with it.

Jets of gas streaming through interstellar space are a common sight in stellar nurseries. Astronomers have found hundreds of jets in star-forming regions, including dozens in the Orion and Taurus-Auriga clouds. These jets travel at supersonic speeds, generating a shock wave where the leading edge hits the surrounding interstellar gas. Heat generated at the shock produces a glowing nebulous patch of gas, typically with a bullet-shaped or conical profile. These nebulae are called

Herbig-Haro objects, after the two astronomers, George Herbig and Guillermo Haro, who first studied them in detail in the 1940s and 1950s (Figure 7.4). At first, Herbig-Haro objects were thought to harbor protostars, but, in the 1980s, their true nature became clear. Although most of the streaming gas is not visible in ordinary light, it can often be tracked along its whole length because the gas emits very short wavelength radio waves. In many cases, a pair of jets moving in opposite directions can be traced back to a single source, even if the source itself is mostly hidden.

After about 100,000 years, much of the original envelope of material surrounding an infant star has collapsed onto the star or the disk surrounding it. The star finally becomes visible from the outside. Today, we

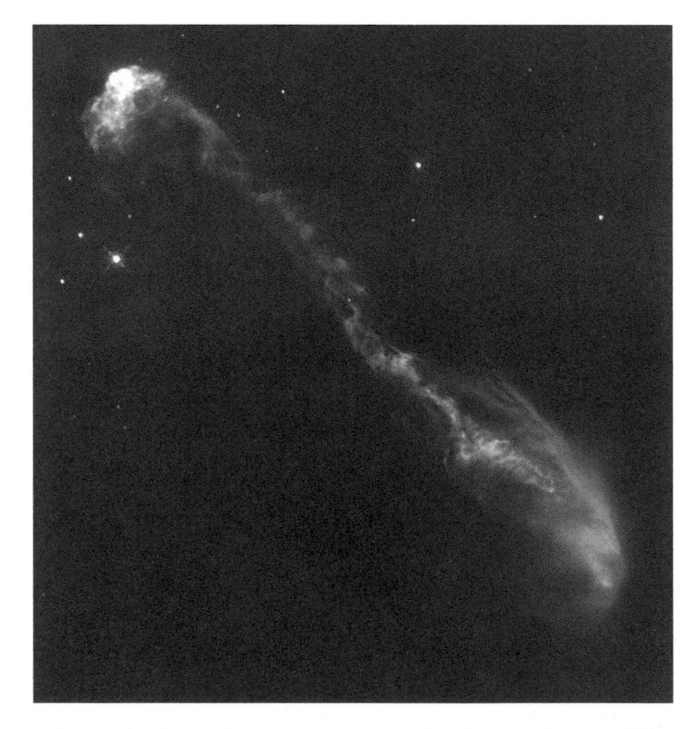

Figure 7.4. The Herbig Haro object HH47, imaged by the Hubble Space Telescope. The object is a jet of glowing gas, half a light-year long, spewed out by a newly forming star that is concealed behind a dark cloud of gas and dust in the lower left corner of the image. (J. Morse [STScI] and NASA)

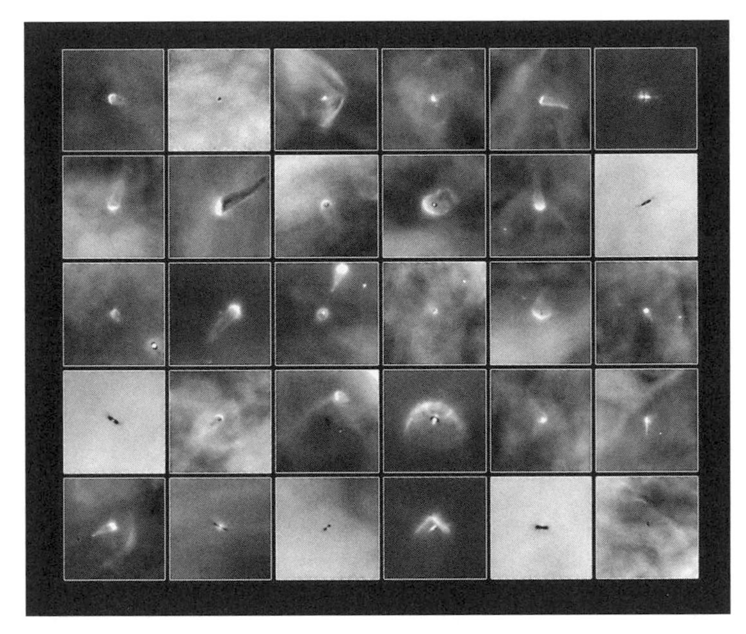

Figure 7.5. Thirty "proplyds" (protoplanetary disks) in the Orion Nebula, imaged by the Hubble Space Telescope. Disks near to the brightest star in the Trapezium cluster glow in the intense radiation from the star. Disks farther away are detected as dark silhouettes against the background of the bright nebula. (NASA/ESA and L. Ricci [ESO])

see many such stars in Orion. In 1994, the Orion Nebula was one of the first targets of the Hubble Space Telescope after it was fitted with corrective optics to fix a fault in its original mirror. Astronomer Robert O'Dell used Hubble's unique clarity of vision to look at 110 young stars in the nebula. He found that 56 of these stars are surrounded by dark, dusty disks, and coined the term "proplyds" for these objects—a shortened form of protoplanetary disk. The images from Orion were compelling evidence that young stars are accompanied by disks as a natural part of star formation. Many of the disks in Orion have sharp edges and are surrounded by glowing shells of gas—clear signs that the outer parts of the disks are being eroded away by ultraviolet radiation from bright nearby stars (Figure 7.5).

The magnetic energy that drives jets of gas from newly forming stars also produces intense flares, similar to the solar flares that occur today

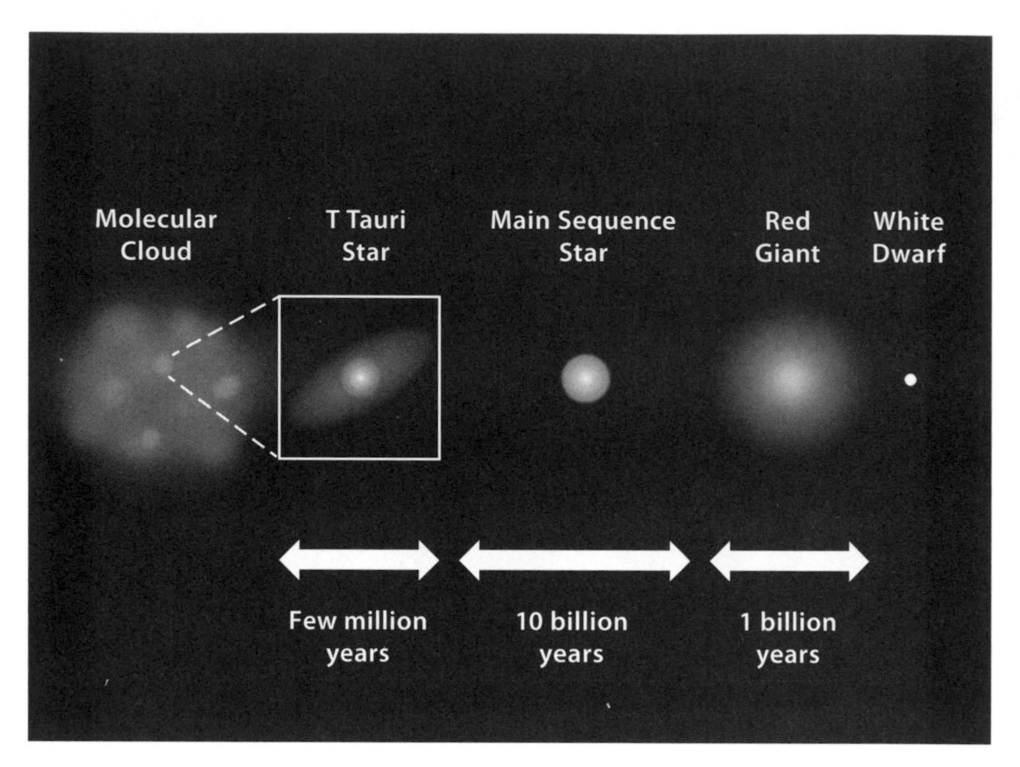

Figure 7.6. The main stages in the evolution of the Sun from its formation to becoming a white dwarf.

but on a much grander scale. Other signs of the violent activity around young stars are powerful blasts of X-rays and radio waves, and temporary outbursts of light as material periodically falls onto a star from the surrounding disk. The Taurus-Auriga region contains dozens of these tempestuous young stars, adjusting to their changing status. One of these, T Tauri, has given its name to this whole category of stars.

T Tauri stars gradually contract and grow denser over time. Material from their protoplanetary disks flows inward onto the star, forms into planets, or is photoevaporated away into space. After a few million years, most disks are gone. Soon afterward, temperatures in the center of the star reach 6 million degrees, and ordinary hydrogen begins to fuse into helium. The violent activity associated with stellar youth ceases and the object settles into a long life as a mature star.

THE SOLAR SYSTEM'S BIRTH ENVIRONMENT

The Sun probably formed, as most stars do, as a member of a cluster within a molecular cloud, but it would not have been in its cluster for long. Like fledglings leaving a nest, stars go their separate ways, and clusters typically disperse in 10 million years or less. Only about 1 percent of stars remain in their birth clusters for longer than 100 million years, and even these stars drift away in the end. Once the gas in a molecular cloud disperses, there is not enough mass left for gravity to hold a cluster together. Computer simulations show that stars mill about within the cluster until they pass close to another member. As two stars approach each other, their mutual gravitational acceleration acts like a slingshot, ejecting one or both stars from the cluster and out into the surrounding galaxy. Sooner or later every star suffers a similar fate and the cluster dissolves.

Did the Sun form in a large cluster alongside massive stars like those in the Orion Nebula, or did it have a more humble origin in a modest cluster like the Taurus-Auriga region? And how did the Sun's particular birth environment shape the solar system? It may seem impossible to answer these questions more than 4.5 billion years later, but some clues remain if you know where to look.

One clue to the Sun's birth environment could be Sedna, a small icy world that lies far beyond Neptune. Sedna moves on a highly elongated orbit ranging between 76 and 960 AU from the Sun, and we will say more about this in Chapter 14. How it got onto such an orbit is something of a mystery. Sedna almost certainly formed closer to the Sun than it is today, either in the Kuiper belt or between the orbits of the giant planets. However, the gravitational pull of the Sun and the planets could not have moved Sedna from its initial orbit onto its current path. Sedna must have felt the pull of another object outside the solar system as well.

Today, fewer than a dozen stars lie within 10 light-years of the Sun, but the situation could have been very different long ago. If the Sun formed in a large cluster, it would have had many close neighbors—the Trapezium cluster in Orion has about 2,000 stars crammed into a region only 20 light-years across. Several stars could have passed close to the Sun before its cluster dispersed. Calculations suggest that a combination of planetary perturbations and the gravitational tugs of passing stars would have populated the entire Oort cloud with small bodies plucked from the solar system, while leaving a few objects stranded on Sedna-like orbits. The stars couldn't have passed closer than 100 to 200 AU from the Sun, however, otherwise their gravity would have left a clear imprint on the orbits of the planets, which we don't see.

Unfortunately, these factors by themselves don't place particularly strong constraints on the nature of the Sun's birth cluster, according to recent research. A second line of evidence comes from the mixture of materials we see in the solar system today. In Chapter 5, we saw how the early solar system contained short-lived radioactive isotopes whose daughter products are found in meteorites. Some of these decay products, such as iron-60, could have formed only in a supernova explosion. This suggests that a supernova happened nearby around the time that the solar system was forming, and that material from the supernova was injected into the Sun's molecular cloud core or its protoplanetary disk.

The progenitor star, a member of the same cluster as the Sun, probably had a mass at least 25 times larger than that of the Sun. Such a massive star would have lived for only a few million years—short enough for the star to explode while the Sun's planetary system was still forming, especially if the massive star formed a little earlier than the Sun. At least one such massive star should have existed nearby if the Sun formed in a large cluster. We can even tell how far away the supernova was. If the explosion were too far away, its radioactive material would have been greatly diluted by the time it reached the solar system, and we wouldn't see its remains today. The supernova couldn't have been too close otherwise it would have disrupted the Sun's protoplanetary disk. Balancing these factors suggests a distance of about two-thirds of a light-year.

Analyzing the numbers in detail, it seems likely that the Sun formed in a cluster containing at least 1,000 stars, possibly many more. Such a cluster would have contained several massive stars, at least one of which became a supernova. During their brief lifetimes, these massive stars would have produced prodigious amounts of ultraviolet radiation. We have already seen how ultraviolet light from massive stars is scouring away the dusty cores in the Eagle Nebula and the proplyds in Orion. In the same way, ultraviolet light might have pared away the outer regions of the Sun's protoplanetary disk. This would explain why the Sun's planets extend out to only 30 AU and why the Kuiper belt has a sharp outer edge only a little beyond this.

Ultraviolet radiation may have left behind a subtler signature as well. When ultraviolet light breaks apart molecules of carbon monoxide in a star's molecular cloud core or its protoplanetary disk, it does so in a way that tends to destroy molecules containing rare isotopes of carbon and oxygen. Some of the liberated oxygen atoms apparently made their way into the inner solar system, which would explain the highly unusual mixture of oxygen isotopes that we see in some meteorites today.

Some astronomers are skeptical that a supernova happened close to the solar system just as it was forming. An alternative idea is that the Sun formed where expanding bubbles of gas produced by several supernovae overlapped, causing a dense molecular cloud to collapse and quickly spawn stars. While the precise conditions remain uncertain, the weight of evidence suggests the Sun began life in a grand stellar cluster like that in Orion rather than the more humble variety we see in Taurus and Auriga.

ESSENTIAL INGREDIENTS

In Chapter 6, we saw how the mixture of elements available to form new stars has been enriched over time. The main ingredients have always been hydrogen and helium, but mature stars have gradually added heavier elements to the interstellar medium as they explode or expel their outer layers into space. These heavier elements are essential for building planetary systems. Rocky planets like Earth are largely composed of oxygen, silicon, iron, and magnesium. Gas-rich planets like Jupiter probably accumulated around a solid core of rock and ice, and also needed heavy elements to begin the formation process. Logically, it should be easier to build planets if a star, and therefore the material surrounding it, is relatively rich in such elements. The ancient red giant star HE0107-5340, which we met earlier, has 200,000 times less of the heavier elements than the Sun for example. Stars as old as HE0107-5340 are very unlikely to have planets because their birth clouds never possessed enough of the essential raw materials. Planet formation may turn out to be a relatively recent phenomenon in our galaxy.

When the Sun was born, debris from earlier generations of stars had already been accumulating in the Milky Way for billions of years. The necessary ingredients to form a star, a protoplanetary disk, and planets were present, but these were only the first steps on the road to the modern solar system. In the next chapter, we will see how a system of planets grew out of the material in the Sun's protoplanetary disk.

EIGHT

NURSERY FOR PLANETS

The solar system has a decidedly two-dimensional aspect to it. The orbits of the eight major planets all lie in almost the same plane, deviating by no more than 7 degrees. Bodies in the asteroid belt and the Kuiper belt stray a little further afield, but these belts are arranged like flattened donuts, aligned with the same plane as the planets. The Jupiter-family comets largely follow the same pattern. Only the Oort cloud and the long-period comets have a truly spherical arrangement.

Kant and Laplace noted the planar nature of the solar system more than two centuries ago. Both used this as the basis for their nebular theories in which the solar system grew out of a flattened disk of matter. In the previous chapter, we saw how young stars like those in the constellation Orion are often surrounded by disk-shaped clouds of gas and dust. Astronomers quickly dubbed these "protoplanetary" disks, assuming that they will form planetary systems, like the nebulae envisaged by Kant and Laplace.

AN EXCESS OF INFRARED

The protoplanetary disks in Orion are too far away to see in detail. Pictures taken with the Hubble Space Telescope show us that they are flattened and dusty but not much else. One way to learn more about disks around stars is to the look at the stars' spectra. In the solar system, Earth and the other planets absorb visible light from the Sun and give off this energy again as infrared radiation. If we viewed the solar system from a great distance, the light from the Sun and planets would merge together, seeming to come from a single source. A dusty disk surrounding a star should intercept substantially more of the star's light than a system of planets would. The total mass of dust might be quite small, but its surface area will be large. In the same way, clouds on Earth are relatively insubstantial objects, yet the billions of tiny water droplets in a cloud are enough to block out the light from the Sun. If a star has a disk, its presence should show up as infrared radiation superimposed on the star's spectrum of visible light (Figure 8.1).

In 1983, astronomers got their first chance to view in some detail almost the whole of the sky as it appears in infrared rather than visible light. In January of that year, a Delta rocket launched IRAS, the Infrared Astronomical Satellite, an observatory that would perform the first infrared survey from space. The IRAS telescope was cooled by liquid helium to a few degrees above absolute zero, almost eliminating heat radiation from the telescope and its instruments. This, and the telescope's location beyond Earth's atmosphere, gave IRAS a stunning new view of the universe.

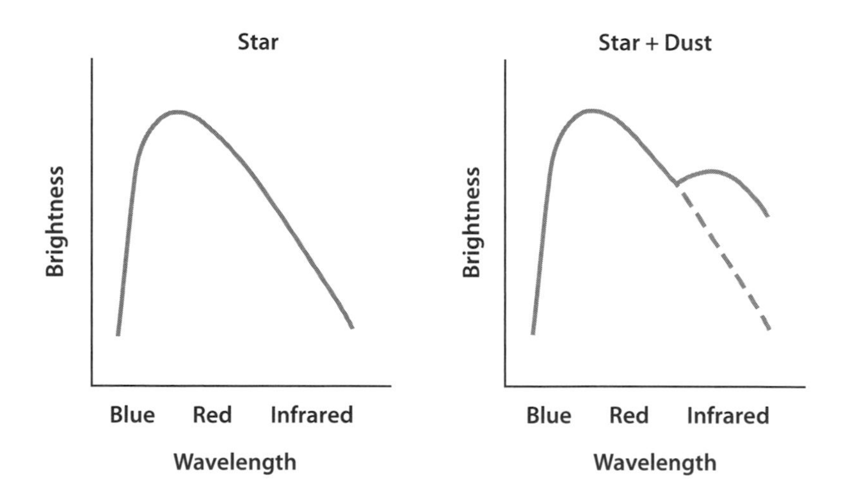

Figure 8.1. Infrared excess. *Left*: A simple schematic graph of the continuous spectrum of a typical star. *Right*: The same spectrum, with the addition of an infrared excess caused by the presence of dust.

IRAS found that some youthful stars emit large amounts of both visible light and infrared radiation. Vega, one of the brightest stars in the night sky, seems to give off at least 10 times more infrared radiation than the star should produce. This "infrared excess" must be coming from dust in orbit around Vega rather than the star itself. IRAS found about 20 stars with infrared excesses, and more recent surveys have found hundreds of others. The wavelength of the infrared excess tells us how hot the dust is and therefore how far from the star it lies. Typically, the dust seen by IRAS is tens or hundreds of astronomical units (AU) from its parent star in each case, very roughly comparable to the distance of the Kuiper belt from the Sun.

A year after IRAS launched, astronomers Bradford Smith and Richard Terrile obtained the first picture of one of these dusty disks. Using a telescope on the ground, Smith and Terrile examined Beta Pictoris, one of the stars known to have an infrared excess. By blocking out the bright glare from the star itself, the astronomers were able to see the faint glow of starlight reflected from dust grains nearby. Beta Pictoris's disk is unusually bright as disks go, and very nearly edge-on when viewed from Earth, which makes it relatively easy to see. In this case, the disk is huge, more than 1,000 AU in radius, 20 times the size of the Kuiper belt in the solar system.

A few dozen other disks have been observed directly with the aid of large telescopes. These include Fomalhaut (the brightest star in the constellation Piscis Austrinus), which has a disk roughly 130 AU in radius, and Epsilon Eridani, one of the nearest stars to the Sun, which is surrounded by a dusty ring at about 50 AU from the star. Beta Pictoris, probably Epsilon Eridani, and possibly Fomalhaut have planetary companions as well, which suggests there is a link between these disks and planet formation (Figure 8.2).

TWO KINDS OF DISKS

Unfortunately, it soon became clear that the disks belonging to Beta Pictoris, Vega, and the other IRAS stars couldn't be true protoplanetary disks. While these disks contain a lot of dust, they probably don't

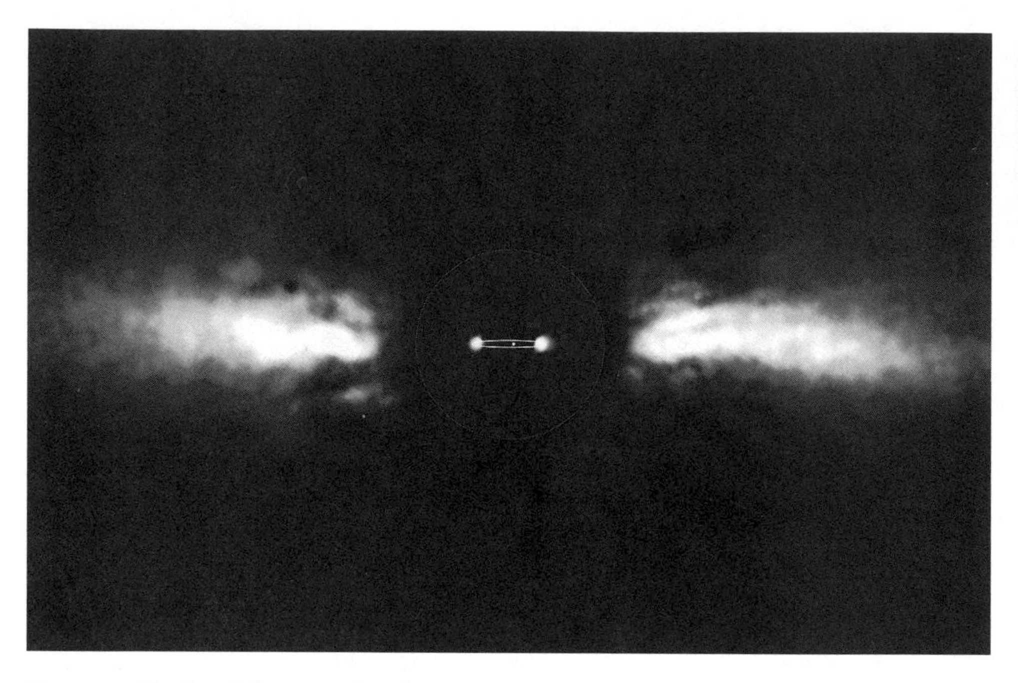

Figure 8.2. The dust disk surrounding the star Beta Pictoris, imaged at the European Southern Observatory, with the possible orbit of the star's known planet superimposed. The orbit has been drawn as a small ellipse for clarity (it is actually almost edge-on), and two actual positions of the planet as recorded in 2003 and 2009 are marked, one either side of the central star. (ESO/A.-M. Lagrange)

contain enough to build a system of planets. Neither do they seem to contain any gas. Stars like the Sun are largely made of hydrogen and helium—elements that would be in the form of gases if they were present in a disk. Since a star and its protoplanetary disk form at the same time, and from the same molecular cloud core, they ought to have a similar composition, but these disks are clearly different. Without hydrogen and helium, the disks seen by IRAS couldn't give rise to gas-rich planets like Jupiter even if they contained enough dust to form rocky planets.

Despite intensive searches, no gas has been found in the disks around Beta Pictoris, Fomalhaut, and Epsilon Eridani. These disks appear to be entirely composed of solid dust grains. Without gas to hold it in place, the dust is highly mobile. Coarse dust grains orbiting each of these stars must frequently collide with other grains at high speeds, breaking apart to form finer dust. Over time, light from the star will alter the orbits of the fine dust grains, causing them to spiral into the star or blowing them into interstellar space. Calculations show that Beta Pictoris and Vega should have lost any primordial dust they possessed long ago. The disks we see today must be made of second-generation dust, formed recently during collisions between asteroids or escaping from comets. Astronomers now call these "debris disks" on the grounds that they are generated by asteroids and comets, which probably represent debris left over from planet formation (Figure 8.3).

While debris disks are not protoplanetary disks, their size, shape, and dusty composition suggest they are close cousins. Stars with debris disks tend to be somewhat older than the T Tauri stars in the Orion nebula and the Taurus-Auriga region that we looked at in the previous chapter. The most obvious explanation is that true protoplanetary disks lose

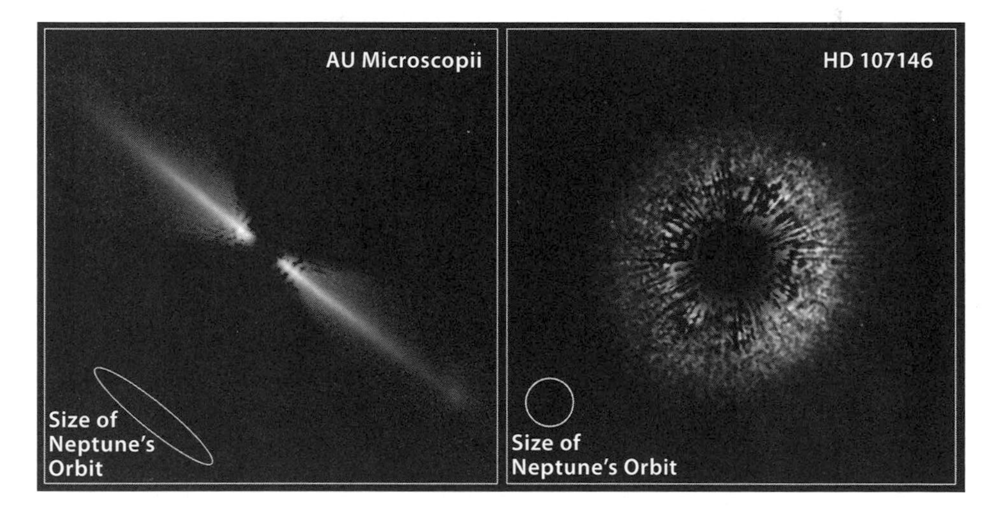

Figure 8.3. Debris disks around two stars, imaged by the Hubble Space Telescope, one almost edge-on and one face-on. The ellipse and circle represent the size of Neptune's orbit to the same scale and in the same orientation as the disks. (AU Microscopii [*left*]: NASA, ESA, J. E. Krist [STScI/JPL], D. R. Ardila [JHU], D. A. Golimowski [JHU], M. Clampin [NASA/GSFC], H. C. Ford [JHU], G. D. Illingworth [UCO-Lick], G. F. Hartig [STScI], and the ACS Science Team; HD107146 [*right*]: NASA, ESA, D. R. Ardila [JHU], D. A. Golimowski [JHU], J. E. Krist [STScI/JPL], M. Clampin [NASA/GSFC], J. P. Williams [UH/IfA], J. P. Blakeslee [JHU], H. C. Ford [JHU], G. F. Hartig [STScI], G. D. Illingworth [UCO-Lick], and the ACS Science Team)

their gas at an early stage and evolve into debris disks as a result. Many stars with debris disks may harbor planets, but it seems that we are a few million years too late to witness the formation of these planets.

Fortunately, there is good evidence that most T Tauri stars have gas-rich disks, even if we cannot see the disks directly. Astronomers have known since the 1940s that the spectra of these very young stars contain emission lines—excess visible and ultraviolet light at certain discrete wavelengths. The brightness of T Tauri stars also varies substantially over time. Both these phenomena are probably caused by hot gas falling onto the star from the inner edge of a gaseous disk. T Tauri stars almost always have infrared excesses as well, which suggests that their disks contain both gas and dust. Unlike most debris disks, the dust orbiting a T Tauri star usually has a broad range of temperatures. This means that dust must be distributed throughout an extensive disk. Typically, the dust in T Tauri disks is spread between a few tenths of an AU from the star out to tens or even hundreds of AU, easily enough to encompass the orbits of all the planets in the solar system.

The strength of emission lines in the spectra of T Tauri stars reveals that gas is falling onto these stars at a rapid rate, typically one Earth mass of gas every few hundred years. Stars up to several million years old are accreting gas at a similar rate. If these stars have been accreting gas this rapidly for so long, their disks must have started with tens of Jupiter masses of material, roughly consistent with the total mass inferred from the amount of dust they contain. This is more than enough material to form the planets in the solar system. These T Tauri disks appear to be true protoplanetary disks like the long-hypothesized solar nebula.

Gas, being essentially transparent, is hard to see in orbit around a T Tauri star. Hydrogen and helium are especially difficult to detect since they generate few features that can be seen in the visible and infrared regions of a protoplanetary disk's spectrum. Other gases, such as carbon monoxide and hydrogen cyanide, are somewhat easier to detect, and these gases have now been found in many protoplanetary disks. Observations made using radio telescopes of some disks show that the gas is rotating around the central star as expected, and that the gas may be turbulent. The spectra of these stars can also be analyzed to identify the materials that make up the dust grains in a disk. Rocky silicates, graphite, ices, complex organic materials—and even tiny diamonds—are commonly seen in protoplanetary disks.

INSIDE THE SOLAR NEBULA

Scientists now have a reasonably clear picture of what the solar nebula must have looked like when the planets were forming thanks to observations of other protoplanetary disks, information gathered from meteorites, and computer models. The solar nebula would have been a roughly symmetrical, disk-shaped cloud of gas and dust centered on the Sun. The disk was thinnest in the center, growing thicker with distance from the Sun. Early in its history, a steady stream of gas and dust fell onto the disk, remnants of the surrounding molecular cloud from which the solar system was born. In the disk itself, gas and dust grains were continuously in motion, rotating around the Sun and slowly flowing inward to be accreted onto the Sun.

The gas in the solar nebula was tenuous, thousands of times thinner than the air in Earth's atmosphere, growing more rarified with increasing distance from the Sun. Most of the gas was hydrogen and helium, with trace amounts of other noble gases, along with water vapor, carbon monoxide, and other molecules. Fine dust grains were everywhere, blocking out much of the visible light from the young Sun, and glowing faintly at infrared wavelengths. Most dust grains were tiny, less than a micron (0.00004 inches) in diameter—much finer than typical household dust particles.

The solar nebula was hottest in the center where solar heating was greatest, and where gas flowing toward the Sun was compressed by the Sun's gravity. The region close to the Sun may have been so hot that everything was vaporized, including rocks and metals. The gas pressure almost everywhere was too low for liquids to be stable, so materials existed as either a solid or a gas. Temperatures declined away from the Sun, allowing more materials to condense as solids. The composition of dust grains varied with position as a result. In the inner nebula, dust consisted of refractory materials such as silicates and metals. Farther out, tar-like organic molecules appeared, then water ice, and finally exotic ices such as solid methane, carbon monoxide, and nitrogen that were stable in only the frigid outermost regions of the nebula.

A few AU from the Sun, temperatures became cool enough for water ice to form. Astronomers call this distance the snow line, and it played an important role in determining the compositions of the planets. Rocky planets like Earth probably formed on the hot side of the snow line, where water existed as a gas, making it hard to hang on to. Today, these planets contain relatively little water as a result. The giant planets and their satellites formed beyond the snow line, so they were able to sweep up large amounts of water ice, and this is reflected in their water-rich compositions.

The gas in the solar nebula was probably turbulent, and this turbulence helped to drive gas inward toward the Sun. Giant swirling eddies millions of kilometers (millions of miles) in diameter spun around slowly over the course of months or years, gradually breaking down into smaller and smaller eddies, ultimately dissipating as heat. Turbulence requires a source of energy to keep it going, otherwise turbulent eddies would die away and the gas begin to flow smoothly. This energy probably came from the rotation of the nebula itself, perhaps due to interactions between the nebula's magnetic field and electrically charged particles, although the details remain unclear.

Chondrules must have been abundant in the solar nebula since they feature so prominently in many meteorites. These roughly millimeter-sized (0.04 inches) rocky particles were clearly heated strongly in the solar nebula at some point. The crystal textures of chondrules tell us they cooled rather slowly, over a period of hours, so they must have formed in large dense swarms, at least hundreds of kilometers (hundreds of miles) across in order to keep their heat for so long. Chondrules have retained some volatile elements such as sulfur that would tend to escape if chondrules were heated for too long. This suggests that temperatures in the nebula were relatively cool where chondrules formed, except during the heating episodes.

Unfortunately, we still don't know what caused the heating events that made chondrules. Some chondrules have fragments of older ones preserved inside them, so heating episodes must have happened more than once, perhaps many times. The wide variety of chondrules and their different ages also suggest that there were multiple heating events, each affecting a portion of the nebula. One theory is that gravitational perturbations from growing planets or dense clumps of gas generated shockwaves in the solar nebula. Any small particles orbiting the Sun would have slowed as they encountered dense gas behind the shockwave. Friction with this gas heated the dust to melting point, and kept it hot until all the heat produced by the shockwave had dissipated. A second idea is that numerous collisions between dust grains built up large electrical voltages rather like those generated in thunderclouds on Earth. When these voltages discharged, they produced lightning that could have heated gas in small regions of the nebula to thousands of degrees, more than enough to melt any dust grains nearby.

Many meteorites contain calcium-aluminum-rich inclusions (CAIs), particles similar in size to chondrules but made of ceramic-like minerals. CAIs must have formed in a very hot environment, although their textures suggest they didn't form the same way as chondrules. The mixture of the minerals in CAIs and the complete absence of volatile constituents suggest that they were among the first solids to condense in the hottest region of the solar nebula close to the Sun. This idea fits with the radiometric ages of CAIs, which show they are the oldest known objects that formed in the solar system. After CAIs formed, they may have been transported away from the Sun by turbulent motions in the gas. Some CAIs were undoubtedly swept up into growing planets, while others would have fallen into the Sun itself, but many ended up in the asteroid belt, where they were joined later by chondrules. Other CAIs were blown into the outer solar system, and one was even found recently in a dust sample returned from comet Wild 2.

GETTING THE DUST TO STICK

Most scientists believe that the formation of the Sun's planets began with fine grains of dust, similar to those seen in protoplanetary disks. These grains coagulated together into larger and larger bodies over time until they reached the size of planets. However, the number of dust grains needed to build a planet like Earth is staggering: a one followed by forty zeroes! This represents a remarkable transformation—all the more so since the solar nebula probably lasted for only a few million years.

Understanding exactly how fine dust grains coagulated into planet-sized objects has been a challenge for many decades. It is impractical to test various theories by building your own planet in a laboratory. However, scientists can examine the early stages of dust coagulation experimentally. One of the biggest problems facing researchers is gravity. Dust grains in the solar nebula moved in free fall about the Sun, so dust coagulation took place under weightless conditions. Finding similar environments on or near Earth is tricky. A few coagulation experiments have been performed aboard NASA's space shuttle in orbit about Earth. In one of these experiments, researchers filled a small container with rarefied gas and tiny dust grains similar in size to dust seen in protoplanetary disks. Once in orbit, astronauts blew a puff of air into the container to separate the dust. In a matter of seconds, the dust grains began to gently bump into one another and stick together, forming long filaments made up of dozens of particles.

In a second test, targets made of fine sand were bombarded with quartz particles roughly 2.5 cm (1 inch) in diameter moving at various speeds. Slow-moving particles embedded themselves in the sand, while high-speed particles bounced off again, loosening some of the sand in the process. These two experiments suggest that dust balls in the solar nebula could grow out of fine dust grains if these grains collided gently. However, energetic collisions would have broken dust balls apart rather than allowing them to grow.

Doing experiments in orbit is a very expensive proposition. Scientists often use cheaper alternatives such as tall "drop towers," where particles fall to Earth from a height of a hundred meters or more (several hundred feet) and effectively feel weightless for a few seconds as a result. Small, cheap rockets that do not fly high enough to achieve orbit can also experience microgravity for several minutes.

One of the most unusual settings for a dust coagulation experiment was NASA's KC135 aircraft, nicknamed the "vomit comet" by those who have ridden it. The pilot of this unfortunate vehicle climbs repeatedly to 10,000 meters (tens of thousands of feet), throttles back the engine and literally falls out of the sky, arresting the dive just in time to avert disaster. During the downward portions of this rollercoaster ride, the occupants of the plane experience microgravity, paying the price during the upswing when everything becomes twice as heavy as normal. Scientists from NASA's Ames Research Center took advantage of this temporary weightless laboratory to examine the growth of dust grains in a small transparent container filled with gas. The dust grains were dispersed throughout the container by a puff of air. Almost immediately they began to stick together, forming aggregates several cm (an inch or more) in size in a matter of seconds. In the solar nebula, dust grains would have been more widely dispersed and growth would have been slower. However, this experiment dramatically demonstrated the willingness of fine dust grains to clump together into large structures.

The results of all these experiments confirm what was found aboard the space shuttle. When dust grains collide at low speeds, they stick to form fluffy dust aggregates. At slightly higher speeds, grains still stick together but the fluffy aggregates become compacted. Somewhat more energetic collisions cause aggregates to bounce off each other instead of sticking, often chipping away individual dust grains in the process. High-speed collisions shatter dusty aggregates into small pieces and rarely lead to growth.

If planets and asteroids simply formed by the gradual accumulation of fine dust grains, we would expect chondritic meteorites to be composed entirely of dust. Individual dust grains would no doubt have become compacted together by collisions and the parent asteroid's gravity, but the grains should have survived and retained their identities. However, this is not what we see in most cases. Chondrites do contain some fine dust grains, but dust is only a minor constituent. These meteorites are mainly composed of chondrules and CAIs, particles that are much larger than the finest dust grains seen in protoplanetary disks.

Collisions between chondrules are very different from those between microscopic dust grains. Experiments show that colliding chondrules always bounce off one another instead of sticking. High-speed collisions often break pieces off the chondrules, so the net effect is a reduction in size rather than growth. Chondrules in meteorites are typically surrounded by rims of fine dust, which they presumably swept up in the solar nebula. If these rims were initially fluffy like the filaments found in the space shuttle sticking experiments, they could change the way chondrules collided. In particular, fluffy dust rims would have absorbed much of the energy of a collision, increasing the chance that chondrules stuck together. Unfortunately, some collisions would inevitably have led to bounces rather than mergers, and the dust rims would have soon became compacted, losing their stickiness. Calculations suggest that clumps of chondrules were unlikely to grow into anything larger than a small boulder even if they had fluffy dust rims.

Clearly, another process apart from simple sticking was needed for chondrules to accumulate into anything larger than a small clump. In fact, as we will see, growth by sticking would inevitably have ground to a halt even in regions that contained only fine dust grains and no chondrules.

THE INFLUENCE OF GAS

Gas in the solar nebula wasn't merely a passive bystander while dust was forming into planets. Instead, it had a strong effect on the solid particles that moved through it. Gas affected small solid objects mainly through drag, what we think of as wind resistance here on Earth. Over time, dust grains and dust balls tended to settle toward the midplane of the nebula, pulled by the Sun's gravity but prevented from falling into the Sun by their orbital motion. Gas drag slowed particles as they fell toward the midplane. Each object moved at a particular steady velocity, in the same way that falling skydivers on Earth reach a maximum speed no matter how far they fall. This speed depended on the size and shape of the particle. Large particles fell the fastest, allowing them to sweep up smaller, slow-moving dust grains en route. By the time they neared the midplane of the nebula, the largest particles were probably about the size of chondrules. Close to the midplane, turbulent motions in the gas arrested their descent and occasionally lofted particles back up again. Particles probably rose and fell repeatedly like ice and water droplets in a thundercloud.

Gas pressure in the solar nebula generally decreased with distance from the Sun. As a result, gas at a particular location felt a slight outward pressure force that partially offset the inward pull of the Sun's gravity. This meant that gas orbited the Sun a little slower than solid objects. Dust grains, chondrules, and other solid particles experienced a headwind as they plowed through the gas while moving around the Sun, and this gradually slowed their forward progress. The continuous drain on their energy caused solid particles to spiral inward toward the Sun. Boulder-sized objects were particularly affected, and objects around 1 meters (a few feet) in diameter would have fallen 1 AU toward the Sun every few hundred years.

Solid particles often became trapped in turbulent eddies in the gas. Small particles were dragged along smoothly within these eddies, while larger objects moved in a more random, haphazard fashion. A combination of these turbulent motions and the inward spiraling of particles toward the Sun meant that particles of different sizes often collided with one another at high speeds—up to 150 km per hour (100 miles per hour). These high-speed collisions were far more likely to cause bouncing or fragmentation than sticking. This, combined with the very short lifetimes of boulder-sized objects due to inward drifting, suggests that—theoretically—growth should inevitably have stalled when objects were somewhere between pebble size and boulder size. Scientists often refer to this problem as the meter-size barrier to growth.

HOW TO BUILD PLANETESIMALS

Scientists have searched for a way that nature might have overcome the meter-size barrier in a single leap, forming asteroids or even planet-sized objects directly from a large population of small particles. Some of the first people to study this problem in detail thought they had found a perfect way to get past the barrier.

In 1969, Soviet scientist Victor Safronov realized that if enough solid particles collected in a thin layer at the midplane of the solar nebula, the combined gravitational attraction of all these particles would make the layer unstable. In a short space of time, particles would spontaneously form into loose clumps, held together solely by gravity. As these clumps contracted further due to their own gravity, they would eventually form solid bodies a few kilometers (a few miles) in diameter. Safronov dubbed these objects planetesimals. He realized immediately that planetesimals would form ideal building blocks for the planets since planetesimals are too large to spiral inward appreciably due to the headwind from the nebula gas, and their gravity would help them to hold on to any fragments broken off during collisions with other objects. The same idea was discovered independently a few years later by American scientists Peter Goldreich and William Ward, and for a time it looked as though the meter-size barrier had been overcome.

Unfortunately, there is a problem with this idea. We now know that the gas in a protoplanetary disk is likely to be turbulent, and the slightest whiff of turbulence is enough to prevent the kind of gravitational instability envisioned by Safronov. As particles settled toward the midplane of the solar nebula, turbulent motions would have stirred them up sufficiently to prevent a thin, gravitationally unstable layer from ever forming.

Scientists have not given up on the idea of jumping across the meter-size barrier, however. Two new theories have been developed recently, both of which embrace turbulence as a fact of nature rather than trying to find ways around it. One idea is that turbulent fluctuations must have occasionally caused many boulder-sized particles to accumulate in one place, albeit temporarily. As boulders piled up, they tended to shield one another from the full effects of the headwind from the nebula gas, in the same way that birds flying in the middle of a flock feel less wind resistance due to the birds around them. Boulders farther out in the nebula continued to spiral inward as before, so they naturally accumulated in regions that already had more boulders than average. Computer simulations suggest that this buildup could have continued until there were enough boulders for gravity to hold them all together, at which point the whole ensemble shrank to form a planetesimal.

The second theory is based on laboratory experiments in which small particles in a turbulent fluid tend to get concentrated in stagnant regions between the swirling eddies. In the solar nebula, this kind of turbulent concentration could have temporarily packed particles much closer together than usual, potentially forming clumps that could have held together and shrunk due to their own gravity, eventually forming planetesimals. Turbulent concentration works only for particles of a special size. What makes this theory particularly promising is that, in the solar nebula, the special-sized particles would have been about 1 mm (0.04 inch) in diameter, precisely the same size as the chondrules that are the major constituent of chondritic meteorites.

At present, we don't know which of these new ideas is correct. Possibly both are wrong, and the true route to planetesimal formation has yet to be discovered. Although the details of planetesimal formation are still unclear, the existence of the Sun's planets and asteroids clearly implies that planetesimals must have formed in the solar nebula. Similarly, the ubiquity of debris disks around young stars tells us that planetesimal formation is a common occurrence elsewhere in the universe.

THE DEMISE OF THE DISK

Compared to stars, protoplanetary disks are short-lived phenomena. Astronomical surveys show that most newly formed stars possess a massive disk of gas and dust. Yet after about 3 million years, roughly half of these disks are gone. Disks, like people, have different life spans. Some pass away young, while others exist into ripe old age. However, very few stars older than 10 million years still have a protoplanetary disk.

Clearly, astronomers cannot watch the complete lifecycle of any individual disk. Instead, we see snapshots of disks at different stages of evolution around different stars, and from this we have to deduce how disks evolve over time. For the first few million years, there seems to be little correlation between the age of a star and the properties of its disk. Disks must slowly lose mass over time as material flows onto the star. At the same time, dust grains in the disk will be accumulating into planetesimals. However, the outward appearance of the disk remains largely unchanged.

Then something happens. Warm dust close to the star seems to disappear, while colder dust remains. Infrared spectra of these disks show that they have a hole in the middle, in the region around the star, where no dust is present. Astronomers call these transition disks because they have properties intermediate between massive, gas-rich protoplanetary disks and low-mass, gas-poor debris disks. Transition disks are relatively rare, which suggests that the transition stage is brief, lasting less than a million years.

The disappearance of the hot dust could be the result of planet formation. As dust coagulates into larger objects, the total exposed surface area decreases, becoming much smaller than for the equivalent mass of fine dust. When this occurs, the star's infrared excess may become too faint to see. If a giant planet forms, its gravitational perturbations will clear a ring-shaped gap in the disk. When several planets are present, their combined gravity could quickly clear away the entire inner disk.

Light from the star may also help form a hole in the disk by photoevaporation. Young stars produce prodigious amounts of ultraviolet radiation. Gas atoms in the surface layers of a disk absorb this radiation, which accelerates the atoms to high speeds. Gas close to the star remains tightly bound to the disk by the star's gravity. However, beyond a few AU from the star, gas is accelerated sufficiently to escape the system altogether, flowing out into space. If the loss rate from photoevaporation ever exceeds the rate at which gas within the disk is flowing toward the star, a gap will open up, rapidly forming a large hole surrounding the star.

Protoplanetary disks can be eroded away from the outside as well, by ultraviolet radiation from nearby massive stars. This appears to be the fate of many of the proplyds in Orion's Trapezium cluster. Low-mass stars are particularly vulnerable because their weaker gravity means that they are less able to hold on to their gas and also because they tend to have less massive disks to begin with.

Not all stars develop transition disks. Some older disks have never formed an inner hole, and their gas and dust seem to be slowly fading away at all distances from the star. This suggests that there are two pathways for disk evolution. Astronomers don't yet know why disks follow one route or the other. It probably depends on how much ultraviolet radiation the star produces and whether large planets form within the lifetime of the disk.

Following either route, protoplanetary disks soon disappear. The remaining gas is removed, falling onto the star, swept up by giant planets, or blown away by photoevaporation. Some of the dust grains accumulate into planetesimals. Most of the remaining grains are lost along with the gas, carried away by gas drag. In the cooler, outer regions of the disk, a small amount of dust remains and is continually replenished by collisions between planetesimals. The protoplanetary disk becomes a debris disk.

The solar nebula provided a sheltered nursery for the early stages of planet formation. The presence of gas damped down the relative speeds of dust grains, limiting disruptive collisions and helping them grow into larger aggregates and chondrules. Turbulent motions within the gas probably helped to concentrate many particles into small regions, allowing their collective gravity to draw the particles together into planetesimals.

If the Sun is like other stars, the solar nebula would have had a fleeting existence, lasting for perhaps 5 or 10 million years, but no longer. However, once planetesimals had formed and reached a certain size, they no longer needed the solar nebula in order to grow larger. These protoplanets were ready to leave their planetary nursery and take the next steps on the road to becoming fully formed planets. In the next chapter, we will see how a few lucky protoplanets in the inner solar nebula eventually became the terrestrial planets.

NINE

WORLDS OF ROCK AND METAL

SISTERS BUT NOT TWINS

Earth and our nearest planetary neighbor, Venus, are superficially similar in several ways. The two planets are nearly the same size and lie in the same region of the solar system. Both bodies are made of rocky materials enveloped by a relatively thin atmosphere. Yet there are obvious differences too. Earth has a magnetic field, but Venus has none. Venus's atmosphere is mainly carbon dioxide, while Earth's consists of nitrogen and oxygen. Earth spins rapidly, once every 24 hours, whereas Venus rotates so slowly that it takes longer to spin on its axis than it does to travel around the Sun.

Even after the advent of large telescopes, Venus's surface remained a mystery for a long time, permanently hidden beneath opaque layers of cloud. Some people speculated that Venus would resemble Earth as it was during an earlier era, a warm tropical world teeming with life. However, a series of spacecraft sent to study Venus by the United States and the Soviet Union in the 1960s and 1970s dispelled the notion of a Venusian paradise. Instead, they found a hellish world, with oven-like temperatures, a crushing atmospheric pressure 100 times that on Earth, and a sky filled with clouds of sulfuric acid.

With these conditions in mind, Soviet engineers designed a hardy breed of spacecraft able to withstand the harsh environment of Venus and explore its surface, if only for a little while. These tank-like vessels were heavily reinforced to withstand the pressure of the atmosphere, and fitted with an elaborate cooling system. One of the most successful of these, Venera 9, embarked for Venus in 1975. On reaching its destination, the craft separated into two parts. An orbiter began to circle Venus, while a lander headed for the planet's surface.

The Venera 9 lander passed safely through Venus's cloud layers, deployed an air brake, and touched down on the ground. Soon afterward, the vessel began to return the first pictures ever taken from the surface of another planet. The images were laboriously transmitted by radio to the orbiter, one pixel at a time, and then beamed back to Earth. They showed a desolate landscape, covered in volcanic rocks, stretching as far as the horizon. The rocks were flat and angular with no sign of the weathering seen in rocks on Earth. Although Venus lies closer to the Sun than does Earth, its surface appeared decidedly gloomy, as its thick clouds reflected away most of the sunlight. Despite these overcast conditions, the surface temperature was more than 460°C (860°F). After 53 minutes, the orbiter moved out of range and the radio link was cut. The lander fell silent, yet it had already shown beyond all doubt that Venus is a truly alien world.

Although Earth and Venus are roughly the same size, they are very different planets. Venus has no oceans, no continents, no magnetic field, no life. Earth and Venus, two planets that could have been so similar are instead a study in contrast. The Sun's other rocky planets, Mars and Mercury, are also very different from Earth, and from each other. In this chapter, we look at how such diverse worlds arose from common beginnings.

THE ERA OF PLANETESIMALS

In the previous chapter, we saw how trillions of tiny dust grains and small particles in the solar nebula clumped together to form a swarm of asteroid-like planetesimals. We also saw how studying other protoplanetary disks sheds some light on the first stages of planetary growth in our own solar system. From this point on, however, other systems offer little to guide us. Planetesimals orbiting other stars are much too small and faint to be seen with telescopes here on Earth. At the same time, planetesimals intercept too little starlight to produce a noticeable infrared excess in a star's spectrum in the way that dust grains do. Planetesimals, the basic building blocks of planets, are essentially invisible to us. To understand how planetesimals evolve into planets, we must examine material left over from planetary growth in the solar system, and make use of computer simulations.

When planetesimals first formed, they were probably loosely bound mixtures of dust grains, chondrules, and other debris that happened to exist in the solar nebula at the time. Today, chondrules and other particles in chondritic meteorites are stuck together so tightly that some effort is required to separate them. This implies that the material inside planetesimals was compressed after these object formed. Planetesimals were too small to become so compacted under their own gravity. However, collisions with other planetesimals would have generated substantial forces, and over time, collisions gradually compressed loosely bound planetesimals until they became solid rock.

Heating caused by radioactive decay profoundly changed the composition of some planetesimals. As they grew hot, volatile materials such as ices and organic tars would have melted and evaporated, ultimately making their way to the surface where they escaped into space. In the more frigid outer regions of the solar nebula, ices were the dominant component of planetesimals. The melting and evaporation of some of this ice probably helped to keep these planetesimals relatively cool and prevented most of the volatile materials from escaping. Some reactions between rock and water took place, but the temperature never rose high enough for rocky materials to melt. In the inner nebula, ices and tars were minor constituents and their cooling effect was limited. These volatile materials escaped almost entirely, leaving behind melted planetesimals composed entirely of rock and metal.

As planetesimals orbited around the Sun, they frequently encountered one another. Some encounters inevitably led to a collision. Computer simulations show that early collisions between planetesimals were energetic enough to fracture and fragment these objects, but generally not violent enough to disperse all the pieces. A planetesimal's gravity was typically strong enough to hold on to many of the fragments, at least at this stage. The net result was the formation of a larger body, together with some escaping debris.

For every collision, there were many near misses. During such close encounters, planetesimals were pulled toward one another by their mutual gravity, only to be dragged apart again as momentum carried each object on its separate way. Although a collision was avoided, these encounters left their mark by altering the orbits of the planetesimals around the Sun.

The combined influence of many encounters left large planetesimals traveling on roughly circular orbits in the plane of the solar nebula. This effect is called dynamical friction. Although no real friction was involved, the combination of many gravitational encounters operated a bit like a frictional force, preventing large planetesimals from straying too far from a simple circular path within the plane of the solar nebula. By contrast, small planetesimals were easily pulled to and fro by their larger neighbors, and they generally ended up moving on inclined and elongated orbits.

Large planetesimals moved on nearly circular, coplanar orbits, and this had important implications for how fast they grew. When two big planetesimals approached each other, the shape of their orbits ensured that they typically moved at a similar speed on almost parallel trajectories. The encounter lasted a long time as a result—often many weeks or months. This allowed plenty of time for the mutual gravitational attraction of the two planetesimals to pull or "focus" their trajectories toward each other, increasing the chance that they would ultimately collide and merge (Figure 9.1).

Small planetesimals increasingly developed inclined, elongated orbits, and they typically approached one another at high speed. Their encounters were brief affairs with little gravitational focusing. As a result, small planetesimals rarely hit one another. When they did collide, their high relative speed often caused a catastrophic breakup rather than growth.

The combination of dynamical friction and gravitational focusing led to a situation called runaway growth: the largest planetesimals rapidly became larger while all the others grew only slowly, if at all.

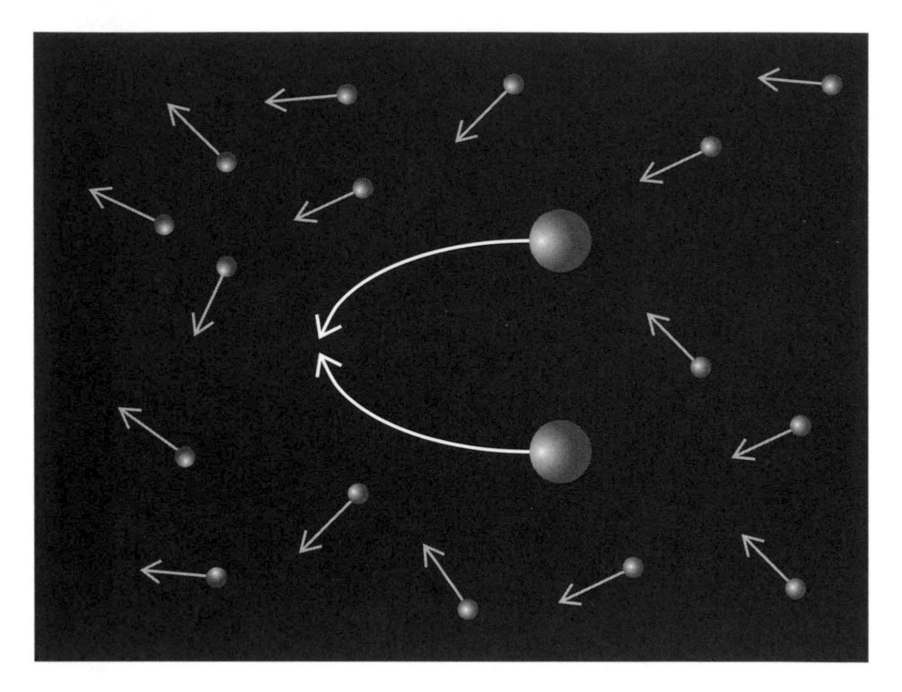

Figure 9.1. Gravitational focusing. Two large planetesimals on almost parallel paths moving at similar speeds had time to be attracted toward each other, increasing the likelihood of a collision.

PLANETARY EMBRYOS TAKE OVER

Unrestrained runaway growth could not last long. Soon, the largest planetesimals became massive enough to perturb most of their smaller brethren onto inclined, elliptical orbits. This made the small objects harder to catch as gravitational focusing grew weaker. Once a large planetesimal had swallowed up all the other large objects nearby, its growth inevitably slowed, as it then had to content itself with pursuing the more elusive small planetesimals.

At this point, a new regime was established. Different regions of the solar nebula each came to be dominated by a single large planetesimal that had absorbed all its rivals. These large bodies are called planetary embryos since at least some of them would ultimately grow into fully formed planets. Dynamical friction meant that embryos moved on almost circular orbits and rarely approached one another. On the few occasions that two embryos met, their mutual gravity soon caused them to merge or made them speed past each other onto orbits that were far apart again.

Each planetary embryo orbited within a swarm of planetesimals, called its feeding zone, swallowing any planetesimals that came too close and growing larger as a result. Planetesimals also collided with one another, but at such high speeds that they rarely merged and often broke apart. Somewhat cynically, astronomers have called this regime "oligarchic growth" after the political system in which a small group of people controls the fate of many (Figure 9.2).

Oligarchic growth was a self-regulating process. If a planetary embryo got greedy and grew too fast, its increasingly strong gravitational tugs quickly stirred up the orbits of planetesimals in its feeding zone. The planetesimals became harder to catch and the embryo's growth was curtailed. Neighboring embryos tended to grow at similar rates as a

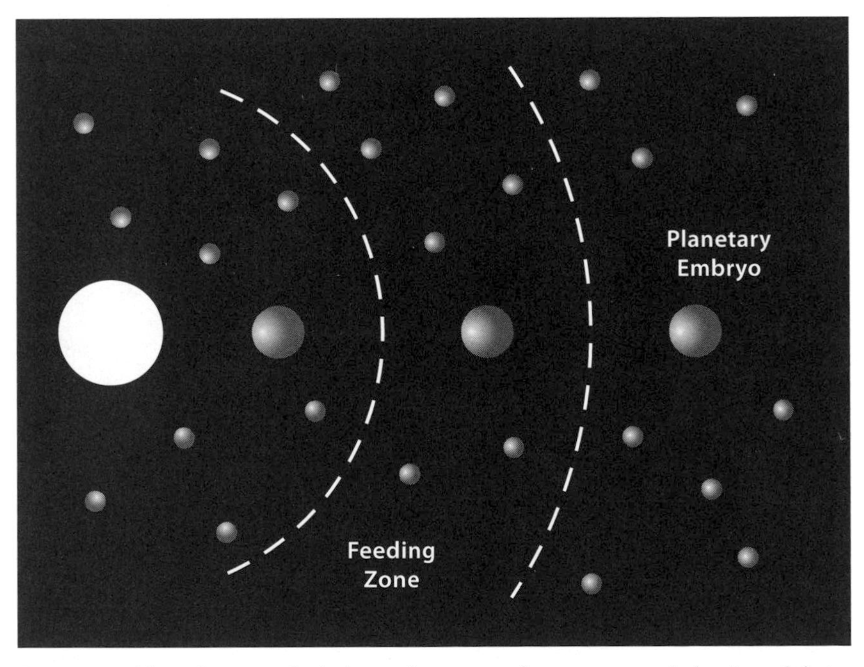

Figure 9.2. Oligarchic growth. As large planetary embryos grew, each dominated their own "feeding zone," becoming larger by absorbing all smaller planetesimals that came close enough.

consequence. Since embryos acquired most of their mass locally from within their own feeding zones, planetary embryos in different parts of the solar nebula came to have different compositions, reflecting the makeup of the planetesimals that formed there.

Computer simulations suggest that oligarchic growth lasted for roughly a million years from the time that planetesimals first appeared in large numbers. As long as oligarchic growth continued, planetary embryos grew larger while the supply of planetesimals dwindled. Gravitational interactions between neighboring embryos became stronger over time, while the dynamical friction caused by planetesimals grew weaker. Eventually, the forces between embryos became so strong that dynamical friction could no longer restrain them. Embryos perturbed one another onto elliptical, inclined orbits, so that they left their feeding zones behind. Gravitational focusing between embryos and planetesimals became much weaker, and oligarchic growth ceased. At this stage, the embryos had swept up roughly half of the total mass of solid material around them. However, the embryos were still only about the size of the Moon or Mars. They still had some way to go before forming planets the size of Earth or Venus.

Planetary growth now entered a prolonged final phase in which long intervals of inactivity were punctuated by brief episodes of unimaginable violence. As the orbits of embryos began to cross one another, encounters became inevitable. However, the inclined and elliptical nature of embryos' orbits meant that gravitational focusing was negligible and collisions happened only occasionally. Near misses continued to change the orbits of embryos, causing them to wander closer to or farther from the Sun in an essentially random fashion. A substantial number of planetesimals still existed at this stage, also moving on inclined, elliptical orbits, and these were slowly swept up by the embryos. The collisions that took place during this phase partially erased the compositional differences established during oligarchic growth.

The strong gravity of the embryos meant that collisions typically led to a merger with relatively little debris escaping back into space. On some occasions, embryos were traveling so fast, or struck each other at such an oblique angle, that they were unable to merge into a single body. In this case, the embryos collided, slid awkwardly past each other, moved apart again, and escaped, exchanging some material in the process. Planetary scientists have dubbed these "hit-and-run" collisions.

Even when they merged, embryos typically hit one another at an angle rather than head-on. The momentum they were carrying was transferred to the newly formed body, making it spin rapidly at an angle determined by the incoming trajectories. When the planets first formed, they probably rotated rapidly, once every few hours. Some may have been spinning on their side, tipped over by the last giant impact they experienced.

Computer simulations suggest that Earth took roughly a hundred million years to reach its current size and clear any remaining planetesimals from its surrounding region. This timescale is roughly in line with estimates based on radiometric dating. The collision rate in the inner solar system gradually declined over time as embryos merged with one another and absorbed the remaining planetesimals or scattered them so that they either fell into the Sun or ended up in the outer solar system.

THE FINAL FOUR

When it was all over, four objects remained—the terrestrial planets we see today. The orbits of these planets don't cross one another, and they are probably far enough apart to prevent any further collisions during the lifetime of the Sun. Table 9.1 shows some of the physical properties of the terrestrial planets together with the Moon for comparison. It is obvious that these objects differ in a number of ways. Earth and Venus are nearly the same size, while Mercury and Mars are significantly smaller. Earth and Mars rotate rapidly while Mercury and Venus spin much more slowly. Earth and Mercury have magnetic fields today while the others do not. The terrestrial planets all have atmospheres, but these range from Venus's atmosphere, which is almost 100 times thicker than Earth's, to that of Mercury, which is so tenuous it hardly deserves the name. The planets' densities also differ, suggesting that their compositions are not all the same. When the effects of compression due to each

Table 9.1. Physical properties of the terrestrial planets

Planet	*Mass (Earth = 1)*	*Average density (water = 1)*	*Uncompressed density (Earth = 1)*	*Rotation period (days)*	*Magnetic field*	*Atmospheric pressure (Earth = 1)*
Mercury	0.06	5.4	1.2	59	Weak	Tenuous
Venus	0.82	5.2	1.0	243	None	93
Earth	1.00	5.5	1.0	1.0	Strong	1.0
Mars	0.11	3.9	0.9	1.0	None	0.006
Moon	0.01	3.3	0.8	27	None	None

planet's gravity are taken into account—yielding the "uncompressed density"—it is clear that Mercury is somewhat denser than the other three terrestrial planets.

The four terrestrial planets all formed in the inner solar nebula from broadly the same population of planetesimals. Each planet has been shaped to various degrees by impacts, gravity, internal heating, and proximity to the Sun, both during and after its formation. Proximity to the giant planets may also have played a role, as we will see. The interplay between these forces has given rise to four very different bodies. In the following sections, we will look at how each planet acquired its unique characteristics, beginning with Earth, the planet we know best.

EARTH

To understand how Earth became the planet we see today, it helps to understand its structure. Geologists are at a distinct disadvantage compared to astronomers in that they can't see directly into Earth the same way astronomers can view the heavens. However, geologists have learned a good deal about the planet's interior through seismometry. When an earthquake occurs, it generates waves that travel through rocks at Earth's surface and in its interior. These waves travel around the globe, where their arrival is recorded by a network of seismometers. Their speed depends on the density of the rock through which they are traveling, so by measuring how long it takes the waves to reach different places, it is possible to measure the density of rocks at different depths inside Earth.

Earth is divided into a series of distinct layers, each with a different density (Figure 9.3). A thin layer at the top constitutes the familiar crust. Underneath the crust lies a denser, rocky mantle that makes up most of the planet's bulk. Rocks from the crust and mantle generally have somewhat different compositions as well as different densities—mantle rocks are richer in magnesium, for example, while crustal rocks tend to contain more silicon. Beneath these rocky layers there is a dense core that takes up about half Earth's diameter and accounts for roughly 30 percent of its mass. Geologists have no direct samples of the core, but its density is so high that it must be 90 percent iron and other metals, and only 10 percent lighter elements.

We can tell that Earth has an iron-rich core in another way too. Certain elements, such as gold, platinum, and iridium, have a strong chemical affinity for iron. These siderophile elements are much less abundant in the rocks that form Earth's crust and mantle than one might expect. It appears that the siderophile elements joined chemically with much of Earth's iron and sank with it to the center, leaving behind a crust and mantle depleted in siderophile elements. In fact, laboratory experiments

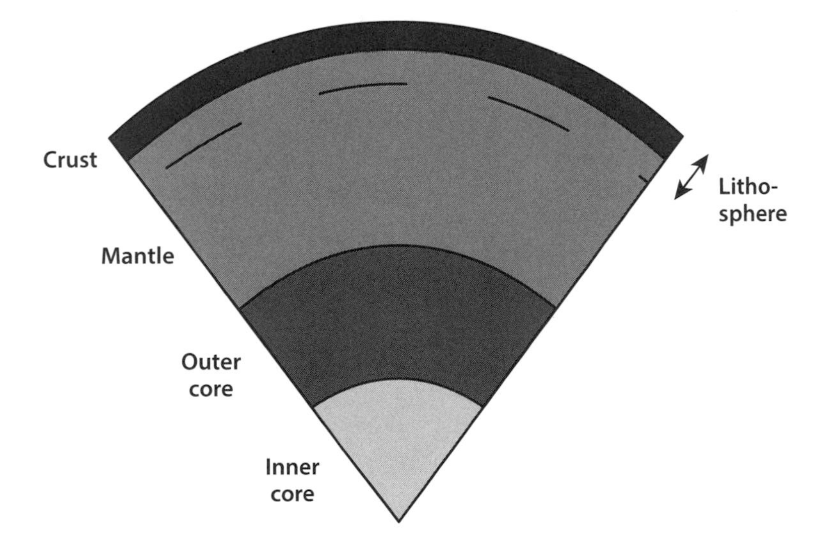

Figure 9.3. The interior structure of Earth.

designed to reproduce conditions deep in Earth's interior show that siderophile elements should be even rarer in the crust and mantle than they actually are. The discrepancy can be explained if Earth acquired a small amount of its bulk after its core had finished forming. This "late veneer," amounting to perhaps half a percent of Earth's mass, brought with it most of the siderophile elements that we find in surface rocks today.

Earth's core probably formed continuously as the planet grew. Core formation required a tremendous amount of heat because iron and other dense materials can separate from rock and sink to the center only in a body that is mostly molten. As we saw in Chapter 5, the parent asteroids of iron meteorites probably melted and formed cores due to heat released by short-lived radioactive isotopes such as aluminum-26. By the time Earth had become a large planetary embryo, most short-lived radioactive isotopes had decayed away, and radioactive heating was becoming much less potent. However, collisions between planetary embryos would have been energetic enough to cause melting in many cases, leading to the formation of a core and rocky mantle. Collisions with other embryos also helped Earth's core to grow as the heavy core from each impacting embryo plunged through Earth's mantle to merge with our planet's own core.

Radiometric dating can tell us approximately when Earth's core formed and, by implication, how long the planet took to grow. The short-lived isotope hafnium-182, with a half-life of 9 million years, is particularly useful. Hafnium is a lithophile element—one that tends to remain in a planet's rocky mantle—while its daughter isotope, tungsten-182, is siderophile and preferentially follows iron to the core of a molten planet. Today, Earth's mantle contains excess tungsten-182, which means that much of the planet's growth and core formation took place before all the hafnium-182 had decayed. The Moon, on the other hand, contains no detectable excess tungsten-182, so it probably formed at least 60 million years after the start of the solar system, toward the end of Earth's growth. We will examine how the Moon formed in detail in the next chapter.

Earth's iron-rich core is believed to have two layers. In the outer core, much of the tremendous heat generated during Earth's formation remains trapped, and temperatures are high enough for iron to be molten. Deeper still lies an inner core where the enormous pressure of the overlying layers of the planet has compressed iron into a solid despite the high temperature. Over time, Earth's interior is gradually cooling, allowing the solid inner core to grow at the expense of the liquid outer core. Heat mostly escapes from the core by convection: hot, low-density portions of the liquid core tend to rise upward, releasing some of their heat as they go, before becoming denser and descending again. The flow of liquid metal in the core generates electric currents that are strong enough to create a substantial magnetic field, a process called the dynamo effect. Earth's rotation tends to align the different convective flows giving the magnetic field a simple north-south configuration, roughly aligned with the planet's spin axis. Today, Earth's magnetic field plays an important role in shielding the atmosphere from erosion by energetic particles emitted by the Sun.

We tend to think of Earth as an ocean planet, but water and ice actually make up only about 0.02 percent of its mass. We may be lucky to possess even this much. Earth probably formed in a part of the solar nebula that was too hot for water ice to be stable. If planetesimals and planetary embryos were to acquire water, they had to capture it in the form of water vapor rather than ice. Recent calculations show that molecules of water vapor would readily stick to dust grains even at the high temperatures seen in the inner solar nebula. Planetesimals that formed from this dust may have been quite wet as a result, at least initially. However, this water was probably lost again soon afterward. Planetesimals and planetary embryos were inevitably heated by the decay of radioactive isotopes and collisions, as we have seen. This heat would have turned water and other volatile materials into gases that readily escaped the weak gravity of planetesimals and disappeared into space.

Most planetesimals that existed in the inner solar nebula may have contained very little water as a result, a conclusion that is supported by the dry nature of meteorites that come from asteroids that were strongly heated in the past. Farther away from the Sun, however, temperatures were low enough for water to freeze as ice. Water ice would have been a major component of planetesimals and planetary embryos that formed in the outer asteroid belt and beyond. Some of this water would have escaped due to radioactive heating. However, the water-rich nature of many carbonaceous chondrite meteorites and comets tells us that planetesimals from the cooler regions of the solar nebula held on to a substantial fraction of their water nonetheless.

It seems likely that Earth acquired much of its water in the form of hydrous planetesimals and planetary embryos that formed farther away from the Sun. During the runaway and oligarchic stages of growth, embryos mostly swept up planetesimals in their own vicinity, so it is unlikely that much exchange of material took place between different regions of the solar nebula. That all changed when oligarchic growth ended. At this point, planetesimals and planetary embryos developed highly elliptical orbits that enabled them to interact and collide over a much wider area of the solar nebula. Once the giant planets Jupiter and Saturn had formed, their gravity altered the orbits of other objects throughout the solar system, causing further mixing.

Inevitably, some water-rich planetesimals and embryos from the cooler regions of the nebula made their way closer to the Sun where they collided with the growing Earth. The energy of the collision released water from the impacting body. Some of this water escaped back into space, but much of it was quickly absorbed into Earth's partially molten rocky mantle. Most of these collisions happened while Earth was growing rapidly, when the core was still separating from the mantle, and some time before the planet acquired its late veneer of siderophile-rich rocks. Later, as the planet cooled over millions of years, steam and other volatile materials in the mantle gradually escaped from volcanoes to form an atmosphere. Eventually, temperatures dropped enough for the steam to condense, forming Earth's oceans. This water played a central role in the emergence and survival of life on Earth, as we will see in Chapter 11.

Today, Earth can be divided into layers in terms of density, but it can also be divided in a somewhat different manner based on the way its rocks behave. The outermost of these layers, called the lithosphere, contains the crust and uppermost mantle. The lithosphere is rigid and divided into roughly a dozen continent-sized pieces called tectonic plates. The bulk of the mantle underneath is solid, but it is not completely rigid, and over millions of years it flows slowly like extremely viscous molasses. Heat deep inside the mantle generates plumes of hot rock that rise upward through the denser surrounding rock until they reach the base of the lithosphere. Here, the plumes spread out sideways and cool, releasing heat from the planet's inner layers.

In some places, the rocks in a tectonic plate can become denser than the mantle beneath. When this happens, the plate descends or "subducts" deep into the mantle, carrying cold material into Earth's interior. As a result, Earth's surface is constantly being destroyed, to be replaced by new crust formed on the ocean floor and elsewhere. Continents are reshaped and relocated as a result of tectonic activity, all driven by heat escaping from Earth's interior. Vigorous plate tectonics, together with the effects of wind and water, have greatly changed Earth's surface over time, largely obliterating all the ancient terrain on the planet.

MERCURY

Mercury, the innermost planet in the solar system, contrasts sharply with Earth. Mercury is a good deal smaller and somewhat denser than our planet when gravitational compression is taken into account. It has almost no atmosphere to speak of, no oceans, and, for the most part, a very ancient surface. Mercury also rotates very slowly, spinning exactly three times for every two orbits around the Sun.

Many of Mercury's properties can be attributed to its small size and its proximity to the Sun. We don't know why Mercury is so much smaller than its nearest neighbors, Venus and Earth. It is conceivable that Mercury's portion of the solar nebula was relatively tenuous, perhaps due to the way matter flowed from the nebula's inner edge onto the Sun. Unfortunately, modern telescopes are unable to resolve other protoplanetary disks clearly enough to see what their inner regions look like, so this is a matter of speculation for now. It could be that the solar nebula contained enough matter to form large planets where Mercury is today, but the growth of planetesimals and planetary embryos was inefficient. Orbital speeds are high this close to the Sun, so collision speeds are typically high as well. Many collisions could have resulted in erosion rather than growth, stunting Mercury as a result.

A third possibility is that Mercury formed somewhere else entirely. Mercury's composition seems to have much in common with enstatite meteorites, which come from the inner asteroid belt. If Mercury formed much farther from the Sun, and later moved to its current location, it may be small for the same reason as Mars, a topic we will return to shortly.

Mercury is the densest of all the planets after allowing for compression due to gravity. An obvious explanation is that it has a very large iron core for its size and a relatively shallow rocky mantle compared to the other terrestrial planets. Data from the Messenger spacecraft, which arrived in orbit around Mercury in March 2011, confirm this interpretation. These data suggest that Mercury's rocky crust and mantle are a mere 400 km (250 miles) thick and overlie a core 2,000 km (1,300 miles) deep.

It was once thought that Mercury has a shallow mantle because intense heating from the Sun caused its upper rocky layers to evaporate into space. This evaporation would have changed Mercury's composition, preferentially removing relatively volatile elements, such as sodium and potassium, while leaving behind refractory elements such as magnesium and aluminum. However, Mercury's extremely tenuous atmosphere contains significant amounts of sodium and potassium. This atmosphere is continually leaking away into space, so it has to be replenished by atoms escaping from surface rocks. It follows that these rocks must still contain substantial amounts of sodium and potassium. Messenger has confirmed that Mercury's surface rocks contain significant quantities of fairly volatile elements, including surprisingly large amounts of sulfur. This seems to rule out solar heating as the cause of the planet's high density.

A more plausible explanation is that the planet suffered a giant impact toward the end of its formation. Computer simulations show that if Mercury's core had already formed at this stage, a high-speed collision with another planetary embryo would have blasted much of Mercury's rocky mantle into space while leaving its core intact. Mercury may have reaccumulated some of this rocky debris at a later date, but much of it could have been swept up by other planetary embryos. An intriguing variation of this hypothesis is that Mercury once collided with a larger planet at an oblique angle, undergoing a hit-and-run collision that stripped away much of Mercury's mantle while the core and the rest of the mantle escaped to form the planet we see today. A giant impact would have generated a great deal of heat, however, and it is still unclear whether this would have allowed Mercury's more volatile constituents to survive.

Mercury is unique among the planets in that its day and year are locked together in a simple ratio. Seen from a fixed point in space, Mercury spins three times on its axis for every two passages around the Sun. This configuration arose due to gravitational perturbations from the Sun. When Mercury first formed, it was probably spinning rapidly as Earth does today. The side of Mercury facing the Sun would have bulged outward due to the Sun's gravity. As the planet rotated, this "tidal bulge" moved slightly ahead of the direction pointing toward the Sun. The Sun's gravity continuously pulled back on the bulge, slowing Mercury's rotation as a result. Had Mercury's orbit been circular, it would have eventually reached a configuration in which the same side of the planet always faces the Sun, in the same way that one side of the Moon always faces Earth. However, Mercury's orbit is quite elliptical, giving rise to the somewhat more complicated configuration we see today.

Mercury's spin axis is almost exactly perpendicular to its orbit—something that has a very low probability of occurring during a planet's formation. It seems likely that the tilt of Mercury's spin axis has been modified over time just like its rotation rate. The culprit is the same in each case—the Sun's gravity—which, together with friction between Mercury's core and mantle, drove the planet's spin axis to its current state. Because of this upright posture, Mercury doesn't experience seasons—unlike Earth, which is tilted by about 23 degrees with respect to its orbit. Mercury's poles are always extremely cold, while at the equator, temperatures fluctuate dramatically between the day and night sides of the planet.

Surprisingly for such a small planet, Mercury has a magnetic field, albeit one that is 100 times weaker than Earth's. Mercury's magnetic field suggests that its iron core is at least partially molten, a notion that is supported by careful observations of the way the planet rotates. Being small, Mercury should have cooled and solidified long ago if it had the same composition as Earth. The fact that Mercury's core is still partially molten suggests it contains large amounts of a light element in addition to iron, lowering the core's freezing point as a result. A plausible candidate for this extra ingredient is sulfur, which is consistent with the sulfur-rich nature of Mercury's surface found by Messenger.

At first glance, Mercury appears rather like the Moon, but there are important differences between the two (Figure 9.4). Mercury's surface is heavily scarred with craters, including some impact basins more than 1,000 km (600 miles) in diameter, but it is not as heavily cratered as the highland regions of the Moon. Large swathes of Mercury's northern hemisphere consist of relatively smooth planes where sheets of molten lava have flooded the terrain and solidified. In places, the ghostly rings of large submerged craters are just perceptible. In addition to its craters, Mercury's surface has a series of steep slopes hundreds of kilometers (hundreds of miles) in length, which apparently formed when the planet shrank significantly as it cooled over time.

Much of Mercury's surface is very ancient, but Messenger has discovered some features that seem to have a more recent origin: thousands of curious pits, dubbed "hollows," that are unique to Mercury. These hollows range in size from a few meters (several feet) to a couple of kilometers (over a mile), and they appear to be younger than the surrounding terrain. Hollows may still be forming at the present time, although how they form remains a mystery. Clearly Mercury can still spring a few surprises on us, and it is not simply a larger version of our Moon as some have supposed.

The ancient volcanic activity on Mercury must have released some gases, but today Mercury lacks any appreciable atmosphere. Several factors affect a planet's ability to hold on to an atmosphere, and all of these seem to work against Mercury. Mercury's weak gravity and high surface temperature allow gases to escape into space more easily than they do from Earth. Its location close to the Sun means that any asteroids and comets that collide with Mercury have been accelerated to high speeds

Figure 9.4. A near global view of Mercury as seen by the Messenger spacecraft on its first flyby. (NASA/Johns Hopkins University Applied Physics Laboratory/Carnegie Institution of Washington)

by the Sun's gravity. High-energy collisions may have blasted away much of Mercury's atmosphere early in its history. Mercury's magnetic field, unlike Earth's, does little to protect it from the intense wind of particles flowing outward from the Sun. Interactions with these particles over the age of the solar system have probably stripped away any atmosphere that survived collisions.

VENUS

Venus illustrates how the evolution of two initially similar objects can follow very different paths. Venus's early evolution was probably similar to Earth's—a lucky planetary embryo that grew larger through a series of giant impacts. Like Earth, Venus would have accumulated some volatile-rich planetesimals and embryos, so initially it may have had a water endowment similar to Earth's. However, solar heating on Venus was more powerful, so more water evaporated into its atmosphere. Water vapor is a strong greenhouse gas, and its presence in the atmosphere would have increased the surface temperature, allowing still more water to evaporate. Eventually, this runaway process gave rise to a thick steam atmosphere. Ultraviolet radiation from the Sun broke apart water molecules in the upper atmosphere into oxygen and hydrogen. The hydrogen, being very light, escaped into space, so that Venus's water was permanently lost. This sequence of events has left its imprint on the tiny amount of water that remains in Venus's atmosphere today. The water contains large amounts of deuterium, which escapes less readily than ordinary hydrogen, telling us that the planet has lost at least enough water to form a global ocean 3 meters (10 feet) deep, and possibly much more.

The loss of its water had profound consequences for Venus. On Earth, water is a key ingredient in chemical reactions that remove carbon dioxide from the atmosphere. Carbon dioxide is another greenhouse gas, and its presence at low levels in Earth's atmosphere has kept our planet at a pleasantly mild temperature for most of its history, as we will see in Chapter 11. Once Venus lost its water, there was no way to remove carbon dioxide from its atmosphere, and this gas built up steadily as it was released from volcanoes. Today, the amount of carbon in Venus's atmosphere is similar to the total carbon in Earth's crust, suggesting that essentially all Venus's carbon has ended up in the atmosphere. With an atmospheric pressure nearly 100 times greater than Earth's and a carbon-dioxide-rich atmosphere, Venus experiences an intense greenhouse effect that keeps its surface at a searing 460°C (860°F).

Lack of water is responsible for another major difference between Venus and Earth. On Earth, water weakens the lithosphere, allowing it to break into plates that move across the planet's surface. Things are very different on Venus. Venus's lithosphere is dry and much stronger than Earth's. It is unable to break into separate plates and instead behaves as a fixed "stagnant lid" sitting on top of the planet's mobile mantle. Venus's surface lacks long, linear mountain chains, midocean ridges, and other features produced by plate tectonics on Earth. Most of Venus's surface lies within 1 km (3,000 feet) of the planet's average height. This is in marked contrast to Earth's surface, which is divided into low-lying ocean floors and high-altitude continents due to plate tectonics.

Our best maps of Venus's surface were made between 1992 and 1994 by the orbiting Magellan spacecraft, which used radar to penetrate the planet's thick clouds and map the surface underneath. Venus's surface is covered in volcanoes and lava plains (Figure 9.5). So far, no volcano has been seen erupting. However, the atmosphere doesn't seem to be in chemical equilibrium, which suggests that it is being continually

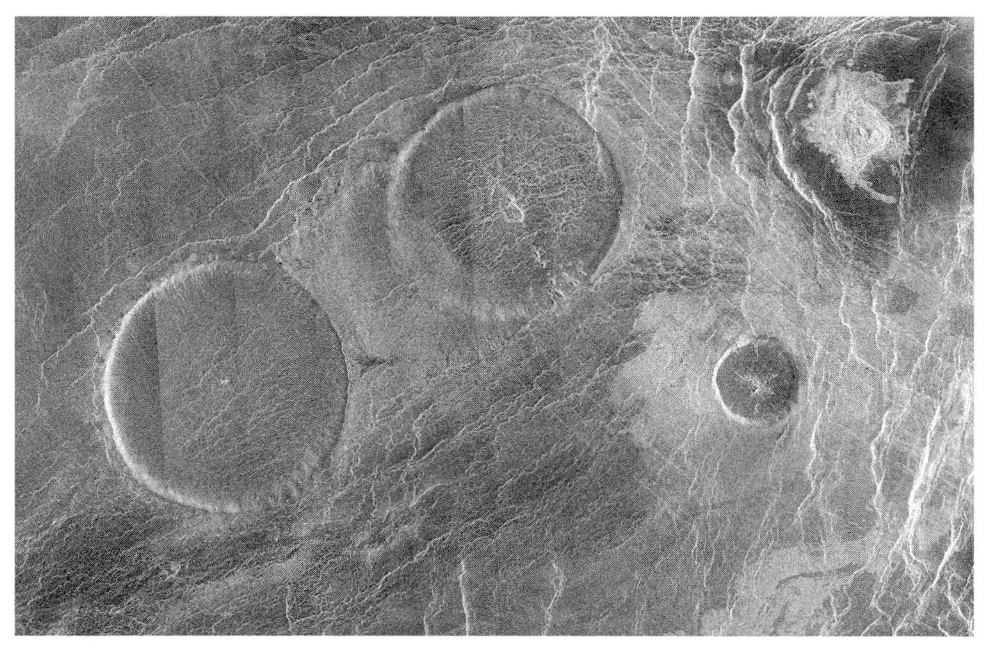

Figure 9.5. "Pancake" domes on Venus. A mosaic of radar images of part of the surface of Venus from the Magellan spacecraft. It shows several flat-topped volcanic domes. The two large ones are about 65 km (39 miles) across and are less than 1 km (0.6 miles) high. This kind of volcanic feature is unique to Venus. (NASA/JPL)

modified by gases released by active volcanoes. The surface is also dotted with numerous circular features, hundreds of kilometers (hundreds of miles) in diameter, called coronae. These appear to be hotspots, where high temperature material from deep within the planet is welling upward and deforming the crust. Earth possesses a few features somewhat similar to these, like the hotspot responsible for forming the Hawaiian islands, but there seem to be many more on Venus.

Around 1,000 impact craters have been found on the surface of Venus. Craters smaller than a few kilometers (miles) in diameter are not seen, which suggests that the planet's atmosphere prevents small asteroids and comets from reaching the ground intact. By counting the number of craters and estimating the frequency of impacts, scientists can tell us how old Venus's surface is. On average, it appears to be only a few hundred million years old, much less than the age of the solar system and broadly comparable to the average age of Earth's surface. This implies that Venus's surface has been modified relatively recently, presumably by volcanic activity. Some scientists have speculated that Venus experienced a catastrophic resurfacing event around half a billion years ago, when pent-up heat from the interior caused the planet's stagnant lid to founder beneath the surface, to be replaced by new material from the mantle underneath. Other less catastrophic scenarios are also consistent with the observed distribution of impact craters, and it is possible that Venus's surface is continually being renewed at a more sedate pace.

Unlike Earth, Venus does not have a magnetic field. The likely explanation is that its core is not convecting vigorously enough. It is unclear why this is the case, but, as with so much else on Venus, it may be linked to the planet's lack of water and the absence of plate tectonics. At present, Venus appears to be releasing its internal heat less efficiently than does Earth, reducing the temperature gradient within the core and preventing a magnetic dynamo from becoming established.

Like Mercury, Venus has had its rotation modified by gravitational interactions with the Sun and by friction between the planet's core and mantle. Venus's dense atmosphere adds a further complication. The side of Venus facing the Sun tends to be the hottest, which increases the atmospheric pressure in this region and causes a flow of atmospheric gas to other parts of the planet. This leaves a low-density region facing the

Sun. Coupled with the planet's rotation, the net effect is the opposite of that caused by the tidal bulge at the planet's surface. Thus, atmospheric heating increases Venus's spin rate while the tidal bulge decreases it. The combination of all these effects over time has given rise to a planet that is spinning backward very slowly around an axis that is almost perpendicular to its orbit.

MARS

Mars, the outermost rocky planet, boasts the solar system's largest volcano (Olympus Mons) and longest canyon (Valles Marineris), as well as the greatest contrast between high and low points on any body—a difference in height of 30 km (100,000 feet). Mars has been comprehensively mapped and studied by orbiting spacecraft and by landers, including four rovers that have traveled across the surface. It continues to be the subject of intense interest and scrutiny because conditions on its surface bear some resemblance to those on Earth, and because Mars is one of the few other places in the solar system that could conceivably support life. However, the differences between Earth and Mars are substantial.

One of the most puzzling things about Mars is its small size compared to Earth. Mars is the sole planet between Earth and the asteroid belt—a considerable expanse of real estate—yet Mars managed to accumulate only enough material to form a planet 9 times less massive than Earth. Computer simulations suggest the most likely explanation is that this part of the solar nebula contained little solid material at the time when Mars was growing. In 2009, planetary scientist Brad Hansen found that he could reproduce the sizes of the four terrestrial planets very well if he assumed all the necessary planetary embryos were initially confined to a ring lying between 0.7 and 1.1 AU from the Sun—entirely interior to the orbit of Mars. Most of these embryos stayed in this region to form Earth and Venus, while a few were perturbed out of the ring to where Mars and Mercury orbit today.

Why would the region between Earth and the asteroid belt contain so little mass? Recently, planetary scientist Kevin Walsh and his collaborators provided a possible explanation. One of the lessons from the discovery of planets orbiting other stars is that planets are highly mobile. Giant planets like Jupiter are prone to migrate through their protoplanetary disk due to gravitational interactions with material in the disk. Typically, a giant migrates toward its star. However, computer simulations show that the migration can be reversed if a second, smaller giant like Saturn lies farther away from the star. Walsh proposed that Jupiter was the first giant to form in the solar system, and that it proceeded to migrate inward through the asteroid belt. By the time Jupiter reached about 1.5 AU from the Sun, Saturn had grown large enough to reverse Jupiter's migration, causing Jupiter to move back out across the asteroid belt to its current location. This scenario has been dubbed the Grand Tack theory, referring to the nautical term for the change in a sailing boat's direction relative to the wind.

The point of this grand tack is that Jupiter's gravity would have wreaked havoc as it migrated through the solar nebula. Most planetesimals and planetary embryos in the asteroid belt would have collided with Jupiter or been tossed into other parts of the solar system. Many objects in the region now occupied by Mars would have met the same fate. The end result was that both regions were depleted of most of the material needed to build planets, leaving only enough mass to form a stunted Mars and even less in the asteroid belt. The in-and-out migration of Jupiter would also have mixed up the orbits of surviving planetesimals, and injected other planetesimals that were rich in water and carbon into the asteroid belt from the outer solar system. This would explain the great diversity of asteroids we see today in addition to the low mass of Mars and the asteroid belt.

One criticism of the Grand Tack model is that it probably requires exquisite timing. Saturn had to form at just the right time: too soon and Jupiter never would have entered the asteroid belt; too late and Jupiter would have disrupted the region where Earth and Venus are today, leaving no large terrestrial planets. Perhaps the solar system was just lucky in this respect, and systems like ours are relatively rare, making Mars something of an oddity. The viability of the Grand Tack model should emerge in the coming years as astronomers start to find large numbers of rocky planets orbiting other stars.

Mars is accompanied by two tiny moons, Phobos and Deimos. They are irregularly shaped, with average diameters of only 22 and 13 km (14 and 8 miles), respectively, and both resemble large boulders deeply pitted with craters. The origin of these moons is still unclear. One theory is that they may be captured asteroids, and they do indeed look very much like asteroids. An alternative idea is that Phobos and Deimos formed in orbit out of material ejected from Mars by an impact.

Like the other terrestrial planets, Mars appears to be differentiated into layers. The orbital motion of Mars's two moons suggests that the planet has a dense core at its center, probably rich in iron. Scientists have reached the same conclusion by studying Martian meteorites, which provide invaluable insights into the planet's composition. Martian meteorites are depleted in siderophile (iron-loving) elements just like rocks on Earth, suggesting that these elements largely reside in an iron-rich core. As with Earth, however, the siderophiles are more abundant than one would expect in this case, so both planets must have acquired some of their mass as a late veneer after the cores had finished forming.

The timing of core formation on Mars can be estimated by measuring the amount of the isotope tungsten-182 in its rocks, in the same way that Earth's core formation has been measured. The Martian meteorites contain substantially more tungsten-182 than do Earth rocks, which means Mars grew faster and its core finished forming at an earlier point in time. Based on its tungsten, Mars must have been almost fully formed within 20 million years of the start of the solar system, and it may have taken as little as 2 million years to form. If the lower figure is correct, it suggests that Mars had reached its current size by the end of the oligarchic growth stage of planet formation, making it more like a leftover planetary embryo than a fully grown planet like Earth.

Mars's surface is clearly divided into two types of terrain. The southern hemisphere is dominated by highlands that are covered in impact craters. The northern hemisphere has a lower elevation and is relatively smooth with few impact craters. Craters on the highlands have a size distribution similar to the craters on the ancient surfaces of the Moon and Mercury, and they presumably formed at around the same time. Craters must have covered the northern hemisphere as well at one point, but this hemisphere was flooded by lava flows after the period of heavy cratering ceased. Several of the craters on Mars are huge, including the Hellas impact basin in the southern hemisphere, which is more than 2,000 km (1,200 miles) in diameter. The collision that formed this crater ejected vast amounts of debris into surrounding areas. This may explain a good deal of the difference between the northern and southern parts of the planet. The difference could also have arisen soon after the planet formed as a result of an even larger impact.

Near the equator of Mars lies a huge, bulging, elevated region. It contains a cluster of large volcanoes, including Olympus Mons, which is almost 30 km (100,000 feet) high and the biggest volcano in the solar system. These volcanoes appear to be several billion years old and lie over hotspots where material from the planet's mantle is welling upward toward the surface. Although hotspots exist on Earth, the motion of plates across Earth's surface means that the volcanoes on top of them never grow very large. On Mars, volcanoes appear to remain stationary, sitting atop the same hotspot where they become ever larger. Some of the lava flows around the Martian volcanoes have very few impact craters, suggesting that these surfaces are young. It's a sign that the volcanoes are probably still active. Many Martian meteorites also seem to come from volcanic regions that were active in the relatively recent past.

Mars has no global magnetic field today, but it clearly had one earlier in its history. Spacecraft orbiting Mars have found that rocks in some regions of the planet's crust are strongly magnetized. These rocks formed when iron-bearing minerals in molten lava aligned themselves with the planet's magnetic field as the lava solidified. Magnetized rocks are seen only in ancient, heavily-cratered areas of the planet and in the oldest of the Martian meteorites. This means that the magnetic field disappeared early in Mars's history, roughly 4 billion years ago, presumably because Mars cooled to the point at which it could no longer maintain a dynamo in its core.

Like that of Venus, Mars's lithosphere consists of a single piece rather than the multiple plates we see on Earth. It is unclear whether plate tectonics once operated on Mars. Intriguingly, the magnetized regions of the crust tend to lie along parallel stripes, which suggests that the ancient surface was spreading outward from a single location, somewhat like the spreading of the ocean floor seen at midocean ridges on Earth that we will examine in Chapter 11. The early stirrings of plate tectonics might also explain some of the difference between the northern and southern hemispheres of Mars. However, it appears that Mars's small size, and corresponding rapid loss of heat, prevented plate tectonics from becoming firmly established. The surface has remained static for most of the planet's history.

Like Earth, Mars has white polar caps at its poles, although these are composed of a mixture of water ice and frozen carbon dioxide rather than pure ice as on Earth. Mars's polar caps currently contain enough water ice to form a global ocean several tens of meters (100 feet) thick. Significant amounts of ice also exist just below the surface at high latitudes away from the equator. In a few places, there is evidence that small amounts of water have flowed on the surface in the recent past. These episodes must have been brief however—liquid water is almost never stable on the surface of Mars today, rapidly freezing or evaporating due to the low atmospheric pressure.

However, there are many signs that large amounts of water existed on Mars's surface earlier in its history (Figure 9.6). Networks of valleys, somewhat similar to those on Earth, are seen in many older parts of the surface. Water seems to have flowed through these valleys for extended periods of time judging by the channels cut into the surrounding rock. A different kind of valley is seen on some younger areas. These appear to have formed during huge flash floods, perhaps by eruptions of water from beneath the surface or the sudden melting of ice (Figure 9.6). In a few places, we see minerals on the surface that must have formed when salty lakes or small seas slowly evaporated away. For instance, in 2011, the Mars rover Opportunity discovered a mineral vein made of gypsum, a hydrated form of calcium sulfate, which was almost certainly deposited by water.

Mars's greater distance from the Sun and its thin atmosphere mean that its surface is generally much colder than Earth's today. For much of Mars's history, the appearance of surface water was probably a transient phenomenon, caused when small regions were heated temporarily by impacts or volcanic activity. However, it is possible that Mars was

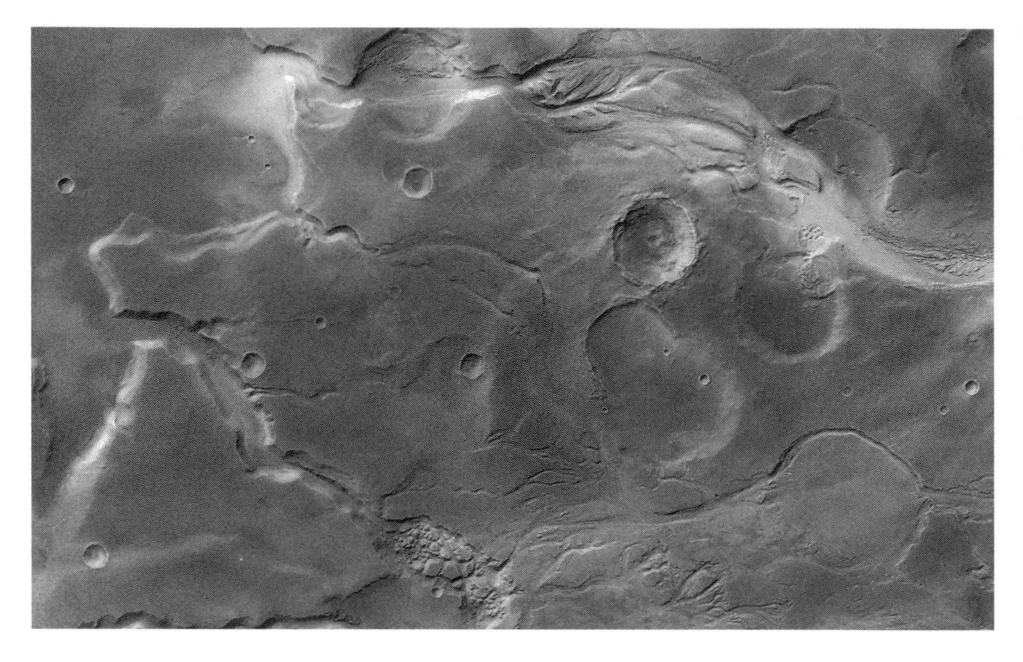

Figure 9.6. The Mangala Valles region on Mars. This system of channels is thought to have been created when liquid water flooded across the surface, probably from under the surface. This image was taken by ESA's Mars Express spacecraft. (ESA/DLR/FU [G. Neukum])

generally warmer early in its history, and this in turn means that it must have had a thicker atmosphere, rich in greenhouse gases such as carbon dioxide and methane.

Volcanoes would have released large amounts of gas early in Mars's history, possibly forming a thick atmosphere. However, the mixture of isotopes we see today suggests that most of these gases have escaped into space. Like Mercury's, Mars's gravity is relatively weak due to its small size, and impacts may have blasted much of its atmosphere into space. Intense ultraviolet radiation from the young Sun and interactions with the solar wind also stripped away much of the planet's atmosphere, especially after Mars's magnetic field disappeared. If Mars once had a thick atmosphere, it seems likely that it vanished within a billion years of the planet's formation. Since then, the absence of plate tectonics and lack of efficient recycling of Mars's crust means that there has been too little volcanic activity to replenish the atmosphere.

Today, Mars's thin atmosphere does little to warm the planet. The surface is almost permanently frozen, dry, and dusty. With no ozone layer to hinder it, ultraviolet radiation from the Sun penetrates all the way to the ground, where it generates highly oxidizing chemicals and destroys organic materials. In these conditions, organisms cannot survive at the surface today. It is possible that life exists underground, but so far we have found no evidence of it. Of all the other terrestrial planets, Mars is the only one where we might find signs of life, either current or past, yet Mars's small size and particular history make it a poor alternative habitat compared to our own planet.

The Sun's four rocky planets have diverse histories that have been shaped by collisions, gravity, radioactive heating, and interactions with the Sun and giant planets. The inner solar system also contains a fifth large, rocky body that has much in common with these planets and yet differs from all of them. In the next chapter, we will see how this body—the Moon—came to be.

TEN

THE MAKING OF THE MOON

The Moon is our nearest neighbor in space, but for a long time it was something of an enigma. With the naked eye, we can clearly make out light and dark regions on the Moon's surface. A modest pair of binoculars reveals a complex landscape of mountains, plains, and craters. However, before the dawn of the space age, we knew only half a Moon: the familiar face turned permanently toward our planet. Most of the lunar far side remained out of reach, less accessible to Earth-bound observers than the most distant planet in the solar system. Even the details we can see on the near side of the Moon raised many questions. What is the true nature of the dark regions that early Moon mappers called "seas" (*maria* in Latin). Why is the Moon's surface covered in craters? Is the Moon a miniature version of Earth, or are the two bodies fundamentally different?

The very existence of the Moon, the only large satellite in the inner solar system, is a puzzle. The Moon is sufficiently large that we would think of it as a planet if it traveled around the Sun rather than Earth. Some of the satellites of Jupiter and Saturn are a little larger than the Moon, but they are tiny in comparison to their host planet. The Moon is an exceptionally substantial companion that must have played an important role in Earth's history. How did such a large body end up orbiting our planet?

THE MOON TODAY

Much of what we now know about the Moon comes from space missions, beginning in the 1960s and early 1970s. Six American Apollo missions each landed two astronauts on the surface. Three of the Soviet Union's unmanned Luna spacecraft touched down on the surface and then returned to Earth. After a long gap, lunar exploration resumed in the 1990s, when NASA's Clementine and Lunar Prospector spacecraft went into orbit. Recently, the pace of exploration has increased again, with the European Space Agency, Japan, China, and India, as well as NASA, all sending missions to the Moon.

These spacecraft have made detailed photographic surveys of the lunar surface and measured the Moon's topography, surface composition, and gravitational field. The Apollo and Luna missions also returned 382 kg (842 pounds) of rock to Earth. These samples, together with around 150 meteorites from the Moon, provide a detailed picture of the Moon's chemical makeup. We now know more about the Moon than any other object in the solar system beyond Earth.

The Moon's surface has two distinct types of terrain (Figure 10.1). Most of the surface is relatively bright and heavily cratered, and these

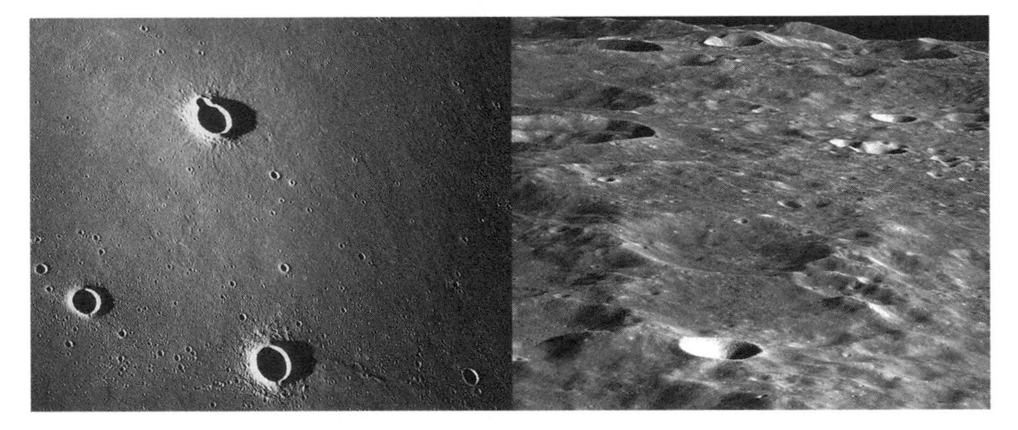

Figure 10.1. The contrast between mare (*left*) and highland (*right*) areas on the Moon. The image on the left is part of Oceanus Procellarum imaged from lunar orbit by Apollo 15 astronauts, that on the right a perspective over highlands on the lunar far side as seen from the Apollo 10 spacecraft. The mare area is relatively smooth, with scattered fresh-looking craters. The highland region is rugged and heavily cratered, with younger craters superimposed on older ones. (NASA/JSC)

regions are traditionally called highlands. The darker mare (singular of *maria*) regions make up about one-third of the near side of the Moon but only a small fraction of the far side. The maria have few craters compared to the highlands and tend to occupy low-lying areas. They are also extremely flat. Samples returned from the maria tell us that they formed when huge lava flows erupted on the lunar surface. This material probably came from deep within the Moon's interior, giving the maria a distinctly darker color and a composition different from that of the surrounding highland rocks.

The lunar surface is dominated by craters and larger circular depressions called basins. Before the space age, scientists debated whether craters were produced by volcanoes or by collisions with asteroids and comets. Close-up pictures taken by spacecraft and samples of lunar rocks soon made it clear that the Moon's craters and basins were formed by impacts. The huge number of craters implies that the Moon has suffered an intense bombardment over its lifetime. Many craters and basins have been partially erased by later impacts. The largest, at 2,500 km (1,600 miles) in diameter, the South Pole–Aitken basin, was firmly identified only in the 1990s. However, the lack of weathering and geological processes over much of the Moon's history means that lunar craters survive for much longer than they do here on Earth.

Impacts have pulverized and churned the upper few meters (several feet) of the lunar surface to form a mixture of fine-grained powder and rock fragments called regolith. Most of the rocks are breccias—broken pieces of rock that have been cemented together by impacts. The regolith is especially deep in highland regions, which have experienced more impacts than the maria. All of the samples returned by the Apollo and Luna missions come from the regolith, and these samples must have been heavily processed since the rocks first formed.

WHAT THE MOON IS MADE OF

At first glance, the Moon seems to be made of the same rocky materials as Earth, but there are important differences. Compared to Earth, the Moon is strongly depleted of the more volatile rock-forming elements such as potassium. Lunar highland rocks contain large amounts of plagioclase (calcium-sodium aluminum silicate)—much more than one would expect if the lunar surface formed by the kind of volcanic processes that occur on Earth. Plagioclase is relatively light compared to most other minerals. This suggests that the upper layers of the Moon were once molten, allowing plagioclase-rich material to float to the surface, and that this material has remained there ever since. The amount of plagioclase in the lunar crust implies that the Moon was once covered by a liquid magma ocean at least several hundred kilometers (a few hundred miles) deep.

Like Earth and the other planets, the Moon is somewhat flattened, bulging outward slightly at the equator due to its rotation. However, the Moon's equatorial bulge is larger than it should be given its current rotation speed. This could be because the Moon's current shape was "frozen in" when it was still hot and partially molten, and rotated more rapidly than it does today.

A major component of rocky materials is oxygen. Oxygen comes in three isotopic varieties, and the relative proportions of each differ from one body in the solar system to another. Earth and Mars have different oxygen isotope ratios, for example, as do most types of meteorite. However, rocks on Earth and the Moon contain identical mixtures. Scientists are still trying to work out the significance of this discovery. It may mean that Earth and the Moon formed out of the same reservoir of material in the same region of the solar system. Alternatively, the two bodies could have exchanged a good deal of material during and after their formation.

Until very recently, the Moon was thought to be almost totally devoid of water and bone dry. However, measurements made by spacecraft have found water ice to a depth of at least 2 meters (6 feet) in about 40 craters near the Moon's north pole. These craters lie permanently in shadow, protecting the ice from solar radiation that would otherwise cause it to evaporate. A new analysis of samples from the Apollo missions also suggests the Moon's interior contains significant amounts of water locked up in minerals, and that the lunar interior may once have been as wet as Earth's.

The Moon has a much lower average density than Earth and the other terrestrial planets. This almost certainly means that the Moon has relatively little iron. It probably contains only 10 percent iron by mass, three times less than Earth does. Most of the Moon's iron is locked up in silicate minerals, although the Moon's gravitational field indicates that a small iron-rich core lies at the Moon's center, containing a few percent of the total mass. Lunar rocks, like terrestrial ones, tend to be strongly deficient in siderophile (iron-loving) elements. This makes sense if these elements tended to bond with iron and sink to the core early in the Moon's history.

Four of the Apollo missions left seismometers on the Moon, allowing scientists to make crude seismic measurements of the interior. Like Earth, the Moon experiences quakes, but moonquakes tend to be much weaker than earthquakes. Seismic data and measurements of the Moon's gravity by orbiting spacecraft show that the Moon has a rocky crust several tens of kilometers (a few tens of miles) thick overlying a denser rocky mantle. The thickness of the crust varies substantially from place to place, typically being thinner on the near side than on the far side. This may explain why the maria are concentrated on the near side of the Moon, since liquid magma would have penetrated the thinner crust more easily here, but there is no clear explanation for the disparity between the Moon's two sides.

THE MOON'S ORBIT

As a whole, the Earth-Moon system contains a surprisingly large amount of angular momentum for its mass. If Earth and Moon were combined into a single body, it would spin once every four hours, much faster than any of the other planets. Ironically, this single detail, which was known long before the advent of space travel, provides one of the strongest constraints on theories for the origin of the Moon, as we will see.

One of the lasting legacies of the Apollo program was a set of small reflectors left behind by astronauts on the surface of the Moon (Figure 10.2). By shining lasers from Earth onto these reflectors and measuring the time it takes for the reflected light to return, scientists have measured the distance between the Moon and Earth with great precision.

Figure 10.2. A reflector set up on the surface of the Moon by Apollo 14 astronauts. (NASA/JSC)

Measurements taken over several decades show that the Moon is slowly receding from Earth, moving away roughly 4 cm (1.6 inches) every year.

The idea that the Moon's orbit might be changing dates back several centuries, to Edmond Halley. One way to test whether this is the case is to examine the timing of eclipses in the historical record and see if they match predictions based on the Moon's current orbit. The first person to do this was the 18th-century English astronomer Richard Dunthorne, who concluded that the Moon is gradually accelerating in its orbit. We now know that most of the effect found by Dunthorne is caused by the gradual slowing of Earth's rotation. However, Dunthorne's measurements firmly established the belief that the Moon's orbit changes over time.

The German philosopher Immanuel Kant realized that gravitational tides would cause the Moon to recede and also alter Earth's rotation. The Moon's gravity causes the side of Earth facing the Moon to bulge outward slightly. This bulge in turn pulls on the Moon. However, Earth's rotation and friction within the planet ensure that the bulge always lies slightly ahead of the direction facing the Moon. Because it is not exactly aligned along the Earth-Moon line, Earth's bulge pulls forward on the Moon, accelerating it in its orbit so that it moves away. At the same time, the Moon pulls back on the bulge, slowing Earth's rotation. This concept was developed further in the middle of the 19th century by Robert Mayer and William Ferrell, who showed that tidal forces must have slowed the Moon's rotation in the past until the same side always faced Earth.

THE FISSION THEORY

George Darwin, whom we first met in Chapter 4, followed these ideas to their logical conclusion. If the Moon is currently moving away from Earth, the two objects should have been much closer together in the past. Perhaps they were once a single fluid body that split in two to form the Earth-Moon system. Matter close to Earth's equator would have been most likely to have escaped due to the centrifugal force caused by the planet's rotation. However, Darwin realized that centrifugal force alone would not have been enough to eject material even if Earth was spinning every four hours. He suggested instead that the Sun's gravity caused matter at Earth's surface to start oscillating vertically. As these oscillations grew larger over time, some material would eventually have moved far enough from Earth to escape into space. According to this "fission theory," Earth would have split into the planet we see today and the Moon.

Soon after Darwin published his theory in 1879, geologist Osmond Fischer added another suggestion. If material had escaped from Earth, it would have left scars that might still be visible today. Fischer argued that the Pacific Ocean basin could be one of these scars, a reminder of dramatic events long ago. Harold Jeffreys offered further support, pointing out that the Moon's shape implies that it must have formed and cooled when it was much closer to Earth than it is today.

It was also Jeffreys, some time later, who showed that friction inside Earth would have damped down vertical oscillations long before they became large enough for material to escape into space. This proved to be a fatal blow for Darwin's model, but not for the fission theory itself. The idea was resurrected in the 1960s by Australian Alfred Ringwood, who argued that Earth initially rotated every two hours, half the period estimated by Darwin. Ringwood suggested that when Earth formed it was initially homogeneous. Iron and other metals sank to the center over time, causing Earth to spin faster and faster until centrifugal force began flinging some material from its surface into space.

Ringwood's idea has several plusses. It naturally explains why the Moon contains little iron. It also explains why Earth and the Moon have identical oxygen isotope ratios: the ratios are the same because the two bodies are made of the same material. Unfortunately, Ringwood's theory also has an inescapable flaw. If Earth once rotated every two hours, the Earth-Moon system today would possess more angular momentum than it actually does. There is no known mechanism that could have removed so much spin over the age of the solar system.

THE CAPTURE HYPOTHESIS

The apparent failure of the fission theory spurred scientists to develop alternative models. Advocates of the "capture hypothesis" suggested that the Moon formed separately at another location in the solar system and was later captured into orbit around Earth. Some satellites of the giant planets, such as Triton, orbit in the opposite direction to their planet's rotation. These satellites surely must have been captured, so there is a precedent for capture elsewhere in the solar system. If the Moon formed far from Earth before it was captured, it would explain why Earth and the Moon have different compositions, although not why their oxygen isotopes are identical.

Capturing a satellite is not easy, however. If the Moon once had enough kinetic energy to wander close to Earth in the first place, it would also have had enough energy to escape again. The Moon could have been permanently captured only if it lost some of its energy in the meantime. Tidal interactions, collisions with other material orbiting Earth, or passage through the planet's atmosphere could remove some energy, but not nearly fast enough to allow the Moon to be captured in a single passage by Earth.

For some incoming trajectories, the Moon could have become captured temporarily into an unstable orbit around Earth. Entering such an orbit requires a very particular set of circumstances, but opportunities to escape are also limited, so the Moon would have spent an extended period of time traveling around Earth before the two parted ways again. During this time, various processes could have removed enough of the Moon's energy for it to be captured permanently, in effect shutting off the escape route before the Moon found its way out. It is possible that this is exactly what happened, but the probability of such a series of events is extremely small. For this reason, most scientists find the capture hypothesis deeply unsatisfactory, and it has largely been abandoned.

THE COACCRETION HYPOTHESIS

A second alternative to Darwin's fission model is that the Moon accumulated gradually from material orbiting around Earth at the same time that our planet was forming. Scientists call this the "coaccretion hypothesis." In Chapter 3, we encountered Laplace's theory for the origin of the solar system in which rings of gas were spun off the Sun, later condensing to form the planets. In 1873, the French astronomer Édouard Roche suggested that something similar happened around the young Earth when a ring of gas spun away from the planet and condensed to form the Moon. To many scientists, this seemed like a plausible way to form the satellites of the giant planets, but it was hard to see why Earth alone would have ended up with a single, large Moon in contrast to the other planets.

A more rigorous version of the coaccretion hypothesis was developed in the 1960s by the Soviet scientist Evgenia Ruskol. As we saw in Chapter 9, there were probably millions of asteroid-sized planetesimals orbiting the Sun at the time when the planets were forming, and many of these planetesimals must have passed close to Earth. Ruskol realized that two planetesimals would occasionally collide and break apart while they were in Earth's vicinity. Much of the resulting debris would have become trapped in orbit around the planet, forming a disk. Over time, more passing planetesimals would have collided with the disk, adding to its mass. Material in the disk would naturally have begun to coalesce into larger bodies, perhaps ultimately forming the Moon.

This raises an immediate question: if the Moon and Earth both formed from the same population of planetesimals, why is the Moon so depleted in iron? Aware of this problem, Ruskol modified her theory, suggesting that weak, rocky planetesimals would preferentially be eroded and end up in the disk around Earth, whereas strong iron-rich planetesimals might pass right through the disk. The coaccretion model thus overcame one obstacle, but it ultimately failed for the same reason as the fission hypothesis: detailed calculations showed that coaccretion would have formed an Earth-Moon system with much less angular momentum than the actual system has today.

By the 1960s, it was far from clear whether any proposed model could explain the origin of the Moon. Solving this mystery became one of the principal scientific goals of the Apollo program. In the event, although the Apollo missions provided a wealth of data about the physical and chemical nature of the Moon, it was not enough for scientific opinion to unite behind any of the existing theories. Instead, the data helped pave the way for a completely new idea.

THE GIANT IMPACT HYPOTHESIS

In the mid-1970s, two teams of scientists suggested that Earth was struck by another planet-sized body in its youth, and that the debris from this collision coalesced into the Moon. This came to be known as the "giant impact hypothesis." William Hartmann and Donald Davis proposed that a lunar-mass body collided with Earth after our planet had already differentiated into an iron core and a rocky mantle. The energy of the impact ejected more than a lunar mass of mantle material into orbit, where many of the more volatile elements escaped to space. Subsequently, the debris coalesced into a Moon that was depleted in both iron and volatiles.

Alastair Cameron and William Ward envisaged a similar scenario, but with one important difference. They proposed that the impactor was larger, with a mass comparable to Mars. In this case, the resulting system would have angular momentum similar to that of Earth and the Moon today, thereby solving a key problem that had defeated earlier proposals. Cameron and Ward also looked more carefully at the details of the impact. One potential problem with the impact hypothesis is that solid debris ejected from the collision site is likely to move outward to a maximum altitude and then fall back to Earth again. Cameron and Ward estimated that the impact would have been violent enough to vaporize much of the ejected material. Pressure from the expanding gas could have pushed enough of the remaining fragments into orbit around Earth to form the Moon.

Despite the obvious attractive features of the giant impact model, it received little attention for almost a decade. A breakthrough came in 1984 at a scientific conference in Hawaii dedicated to the subject of the Moon's formation. Several scientific papers were presented there showing that all the classical models were fatally flawed. At the same time, a new generation of computers and calculation methods had made it possible to test the giant impact model in detail. In a series of presentations, scientists showed that a giant impact could explain all the main features of the Moon. By the end of the meeting, the giant impact hypothesis had emerged as the clear favorite.

Almost three decades later, scientific discoveries and state-of-the-art computer simulations have filled in many of the details that were unclear at the time of the Hawaii conference. We now know that the final stage of planet formation was marked by many giant collisions between planetary embryos, so the formation of the Moon in a giant impact is entirely plausible. Heat generated by these collisions also caused embryos to differentiate into iron-rich cores surrounded by rocky mantles, a key requirement of the giant impact scenario.

ENCOUNTER WITH THEIA

According to the giant impact theory, near the end of its growth, Earth was struck by a Mars-sized object that came from somewhere between the orbits of Venus and Mars (Figure 10.3). Scientists call this rogue body Theia after the Greek goddess who was the mother of the Moon. Theia struck the young Earth at a glancing angle. Much of Theia's bulk came to a halt and merged with Earth. Theia's iron core plummeted through Earth's mantle and soon coalesced with Earth's existing core. At the same time, the part of Theia's mantle farthest from Earth was sheared off by the collision and continued on its way. This material, under the influence of Earth's gravity, soon changed direction and went into orbit, where powerful tidal forces ripped it into small pieces. In a matter of hours, Earth swallowed up four-fifths of Theia's bulk, while the remaining fraction formed a close-orbiting disk of debris.

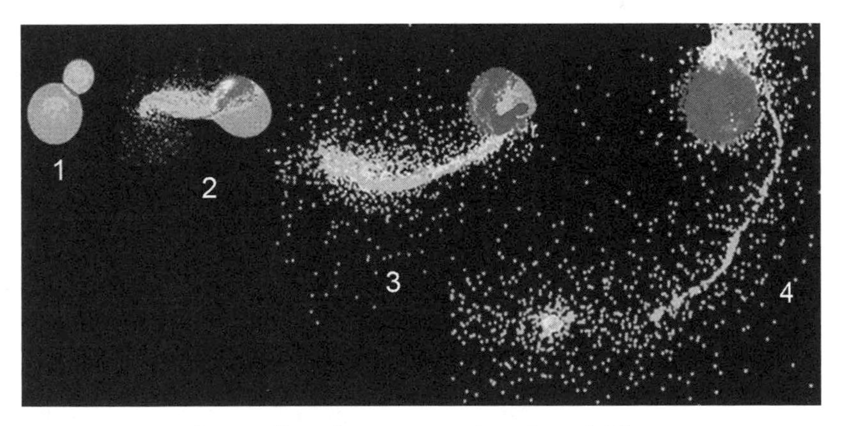

Figure 10.3. A simulation of how the Moon may have formed following a giant impact on Earth. (1) after 6 minutes; (2) after 52 minutes; (3) after 2 hours 9 minutes; (4) after 4 hours 51 minutes. (Adapted from a computer simulation by Robin Canup, Southwest Research Institute)

Most of the material in the disk came from Theia, with a small contribution from Earth itself. Overwhelmingly, the mass came from the mantles of the two bodies and not their cores, explaining why very little metallic iron was present. The energy of the collision generated tremendous amounts of heat, so the disk was composed mainly of boulder-sized droplets of molten rock together with some gas. Some water and other volatile materials evaporated and escaped to space, leaving behind rock enriched in refractory elements but apparently still containing some water.

Following the impact, the disk began to radiate heat into space, gradually cooling over time. Simultaneously, gravitational interactions within the disk caused it to spread apart. Material at the inner edge fell to Earth, while the disk's outer edge moved outward. Over the course of several decades, particles in the disk cooled and solidified until Earth was surrounded by a disk of rocky boulders extending out to several times Earth's radius. Turbulence and gravitational interactions within the disk mixed material from different regions together, and gas moved back and forth between the disk's inner edge and Earth's atmosphere. This kind of mixing probably explains why Earth and the Moon have similar oxygen isotope mixtures today.

The rocky particles orbiting Earth were packed closely together and frequently collided with one another. However, collisions didn't always result in a merger. Close to Earth, tidal forces ripped apart large molten objects and prevented particles from merging together. Beyond about three Earth radii, tides were weak enough to allow particles to merge and grow larger. This distance is called the Roche limit after Édouard Roche, whom we met earlier. As the disk cooled and spread, solid material accumulated beyond the Roche limit and quickly began coalescing into larger bodies.

It is conceivable that a single object swept up all the material in the disk, forming the Moon directly. It is also possible that multiple moonlets formed. However, these moonlets would not have retained their separate identities for long. Tidal forces from Earth and the disk would have driven them outward, with the largest one moving the fastest. A combination of tidal forces and gravitational interactions between moonlets soon caused them to merge or fall back to Earth. Within a few thousand years at most, only a single Moon remained.

EARTH, MOON, AND TIDAL FORCES

The giant impact hypothesis predicts that the Moon formed close to Earth. Today, the Moon is comparatively far away, lying 384,400 km (239,000 miles), or roughly 60 Earth radii from the center of our planet. This change can be attributed to tidal forces that caused the Moon's orbit to expand and at the same time slowed Earth's rotation. Tidally induced evolution would have been very rapid at first, with the Moon's orbit doubling in size in as little as 10,000 years. Within 100 million years, the Moon had moved from just outside the Roche radius to perhaps 20 Earth radii away, one-third of its current distance. At the same time, the Moon's rotation rate slowed rapidly. The current synchronous rotation of the Moon, with one side permanently facing Earth, was probably established at a very early stage.

Tidal evolution slowed down over time as the Moon receded and gravitational interactions between the Moon and Earth grew weaker. The strength of tidal interactions probably depended quite sensitively on the positions of the continents on Earth's surface, since most of the energy dissipation associated with tidal evolution occurs in shallow seas. The Moon's recession rate probably sped up and slowed down as the continents changed position. We can see signs of the long-term change in Earth's rotation in the fossil record. For example, some fossilized seashells contain layers that were laid down once per day. By counting these layers, scientists have found that 350 million years ago there were roughly 400 days in a year, 35 more than at present.

Since the time of George Darwin, scientists have attempted to turn back the clock, calculating the Moon's orbit backward in time to the point when it was close to Earth. These calculations show that the Moon followed an orbit that was tilted to Earth's equator by about 10 degrees soon after it formed. The giant impact hypothesis, together with most other models for the Moon's origin, predict that the Moon's orbit should line up almost exactly with Earth's equator. This unresolved "inclination problem" shows that scientists still do not entirely understand the Moon's tidal evolution. The Moon's orbit may have become tilted by gravitational interactions with material left over in the disk from which it formed. Alternatively, Earth may have experienced another large impact that tilted its spin axis after the Moon formed.

Another consequence of tidal interaction is that the angle between Earth's equator and its orbit around the Sun has increased over time. This angle, called the obliquity, determines the strength of seasonal variations on our planet. As Earth's obliquity has increased, the differences between summer and winter have grown stronger. At the same time, the Moon has played an important role in stabilizing Earth's climate on shorter timescales. Gravitational tugs from the Sun, the Moon, and the planets cause Earth's spin axis to wobble around—or precess—with a period of roughly 26,000 years. These gravitational perturbations also cause Earth's orbital plane to precess, but at a much slower rate. Because these rates are very different, they are essentially separate processes, and this ensures that Earth's obliquity remains almost constant apart from the very slow increase caused by the Moon's recession. Mars, by contrast, has no large satellite, and its orbit and spin axis precess at similar rates. This causes the tilt of Mars to vary dramatically, changing by tens of degrees in only a few million years. These variations lead to substantial climate fluctuations, reflected in the periodic expansion and retreat of Mars's polar ice caps.

Earth's benign configuration will not last forever. In the future, the Moon will continue to recede from Earth, and its gravitational effects will grow weaker. Earth's spin axis will precess more slowly as a result. Sometime in the next 2 billion years, the Moon's influence will become so weak that Earth's obliquity will start to undergo wild variations, as is the case with Mars, producing large swings in the climate. Looking further ahead, if Earth survives the expansion of the Sun during its red giant phase (see Chapter 6), Earth's rotation will slow until one side permanently faces the Moon. At this point, the Moon will stop receding from Earth and begin moving toward our planet at an ever increasing rate. Eventually, the Moon could end its life as it began, in a giant impact.

LATE HEAVY BOMBARDMENT

One of goals of the Apollo missions was to find so-called "genesis rocks" dating back to the formation of the Moon roughly 4.45 billion years ago. Analyzing such rocks would tell us much about the Moon's formation and early history. The oldest lunar rocks brought to Earth by the Apollo missions are indeed of this age, only about 100 million years younger than the solar system, and nearly as old as the Moon itself. These ancient rocks were made when the Moon's magma ocean cooled and solidified. Different minerals formed in stages as the temperature dropped. The first to form contained iron and sank because they were denser than the liquid magma. Crystallization of most of the magma took no more than 100 million years, and the lighter minerals that formed the crust rose to the surface about three-quarters of the way through the process.

Unfortunately, intact genesis rocks proved to be rare in the Apollo samples. Instead, most rocks returned to Earth were the pulverized remains of older rocks that had been substantially modified by impacts. An impact not only breaks rocks into pieces but also melts part of the material. This melting causes minerals to mix together and allows trapped gases to escape, resetting the radiometric clocks used to calculate a rock's age, as we saw in Chapter 4. Impact-melt rocks provide an excellent way to determine the timing of impact events, but not the formation of the Moon itself.

Many of the melts in the Apollo samples have ages clustered around 3.9 billion years, half a billion years younger than the Moon itself. Impact melts older than about 4 billion years are very rare by comparison. Some dated lunar samples can be traced to particular impact basins, indicating when these basins formed. Astronomers can also deduce the order in which basins formed by looking at how material ejected from one basin overlaps another. Taken together, this information tells us that at least half a dozen impact basins formed in an interval of less than 100 million years, a very small fraction of lunar history.

Impact melts are preferentially formed in large collisions, so melt samples tend to date basin-forming impacts. However, the rate at which smaller impacts occurred can be estimated by counting the number of craters in regions with known ages. Some highland regions are saturated in craters—they contain so many craters that additional impacts would have destroyed as many craters as they created. The maria contain few craters by contrast, even though rock samples from mare regions tell us they formed only a few hundred million years after the impact basins. Clearly the frequency of impacts fell dramatically in the interim. Scientists have estimated that the cratering rate declined by at least a factor of 100 between 3.9 billion and 3.0 billion years ago.

Taken together, the cluster of impact melt ages around 3.9 billion years, the paucity of older melts, and the rapid decline in the cratering rate after 3.9 billion years imply there was a brief but dramatic flurry of impacts half a billion years after the Moon formed. This event is referred to as the "late heavy bombardment"—"late" to distinguish it from the even heavier bombardment that earlier accompanied the formation of the planets.

As the bombardment tapered off, volcanism became the dominant process modifying the Moon's appearance. Molten lava flowed across large areas of the surface, flooding impact basins, and explosive eruptions sent material flying over hundreds of kilometers (hundreds of miles). Volcanic activity peaked between 3.5 and 3.0 billion years ago, although some volcanic rocks appear to have formed as recently as 1 billion years ago.

Like the surface of the Moon, the ancient surfaces of Mercury and Mars are covered in impact craters. The ages of these craters are unknown, but they have the same size distribution as those on the lunar highlands. The size of a crater is directly related to the size of the impactor that caused it, so this suggests that Mercury, Mars, and the Moon were all bombarded by the same population of small bodies at the same time. Several types of meteorites contain impact melts with ages that cluster around 3.9 billion years, which suggests that the late heavy bombardment also extended into the asteroid belt. Presumably, Earth and Venus experienced many impacts at the same time, but the craters formed by these collisions have been erased by geological processes.

It is still unclear what caused the late heavy bombardment, although there are a number of theories. Some planetesimals left over from the formation of the planets may have remained in orbit around the Sun for millions of years, eventually colliding with the Moon and planets, but calculations show that most of this material would have disappeared by the time of the late heavy bombardment. A sudden spike in the impact rate could have been caused by the collisional disruption of an asteroid similar to Ceres, as debris from the collision swarmed through the inner solar system. However, the probability of such an event is tiny.

The most likely explanation is that there was a sudden change in the orbits of the planets around 3.9 billion years ago. As we will see in Chapter 14, there is a theory that predicts such a change, which would have altered the orbits of many asteroids and comets, setting some of them on course to collide with the Moon and the planets. This idea is supported by the size distribution of craters on the lunar highlands, which suggests that the impactors came from the asteroid belt.

Although the cause remains uncertain, the scars visible on the surface of the Moon are a reminder of a violent episode early in the history of the solar system. The bombardment that formed the lunar basins almost certainly extended to Earth as well, and it had profound consequences for the development of life on our planet, as we will see in the next chapter.

ELEVEN

EARTH, CRADLE OF LIFE

THE HADEAN ERA

Earth must have been a nightmarish place immediately after the giant impact that formed the Moon. Shock waves raced away from the site of the impact, spreading in all directions. These waves traveled through Earth's interior and converged again on the far side of the planet, blasting away much of the atmosphere into space. The tremendous energy released during the impact melted the upper layers of Earth into a magma ocean, a slushy mixture of molten and solid rock more than 1,000 km (600 miles) deep. If the young Earth had any oceans of water, they would have boiled almost instantly during the collision. In the ensuing hours and days, the planet became enveloped in a dense atmosphere of steam and vaporized rock. Water and other volatile materials continuously moved back and forth between the atmosphere and the magma ocean, dissolving in the hot rock and being transported into the interior.

Temperatures began to fall soon after the impact as Earth radiated heat away into space. Within a few thousand years, rock vapor in the atmosphere had condensed back onto the surface. The magma ocean began to solidify from the inside out. Over the next million years, steam in the atmosphere condensed and rained onto the surface, forming a global ocean of water. After this deluge, Earth's atmosphere consisted of a thick blanket of carbon dioxide with trace amounts of nitrogen and other gases, quite like the atmosphere of Venus today. Chemical reactions between carbon dioxide and water soon began to form vast layers of carbonate rock at the bottom of the ocean, gradually removing carbon dioxide from the atmosphere.

Earth's surface continued to suffer significant impacts as the debris left over from the formation of the planets was swept up, although none of these collisions was as violent as the one that formed the Moon. The largest impacts partially vaporized the ocean, temporarily forming a steam atmosphere. The atmosphere was also bathed in intense ultraviolet radiation from the young Sun, which broke apart water molecules into hydrogen and oxygen. The lighter hydrogen escaped to space, dragging some of the other gas with it. Much of Earth's early atmosphere was lost in this way, although it was continually replenished by gases leaking out of the interior.

Scientists call this earliest phase of Earth's history the Hadean, from the Greek word for the underworld. The Hadean Earth was very different from the planet we see today. This was a time of unimaginable cataclysms interspersed by long periods of relative calm. It was also the period when Earth was transformed from a hellish ball of molten rock into something that began to resemble our world today.

Early Earth consisted of two layers: a dense iron-rich core at the center, surrounded by a thick rocky mantle. Although most of the mantle quickly solidified, it remained hot and flexible, able to flow slowly like molasses. The large amount of heat trapped in the planet's interior caused the mantle to convect slowly. Hot plumes of rock from the deep interior rose toward the surface over millions of years, gradually releasing heat until they became denser than their surroundings, at which point they began to sink again. In this way, heat escaped from Earth's interior much more rapidly than it would have done by thermal conduction alone.

As each plume of hot rock rose through the mantle, the pressure surrounding it decreased. With less compression to keep it solid, some portions of the rock could liquefy. Elements such as sodium, potassium, and calcium that do not fit comfortably into the main rocky minerals preferentially entered the liquid rock or "melt." In places where the melt erupted at the surface as basaltic lava, it was rich in these "incompatible elements." Gradually, Earth acquired a third layer, a thin basaltic crust on the floor of the ocean. In a few places, the basaltic rock became thick enough to poke up above the waves to form the first primitive continents.

As the basalt cooled, it grew denser and eventually became heavier than the mantle below. This configuration was unstable, and in places, the heavy basaltic crust began to sink, or subduct, back into the mantle, where it eventually mixed with rocks deep in Earth's interior. As the basaltic crust subducted, it pulled neighboring material with it, making room for new crust to form and setting up the conditions necessary for plate tectonics—the process of crustal recycling that continues to operate on Earth today. The surface of Earth became divided into roughly a dozen plates moving slowly around the globe at a few centimeters (up to a few inches) per year.

Today, basaltic lava erupts mainly at midocean ridges—long, jagged peaks that rise above the ocean floor for thousands of kilometers (thousands of miles). As lava erupts and solidifies, the new rock is dragged down and away from the ridge by its own weight, and pulled along by material subducting at the edge of the plate, causing the new ocean floor to spread apart and move away from the ridge. Over tens of millions of years, the spreading ocean floor gradually cools and ultimately subducts back into the mantle.

When midocean ridge basalts first form, they react with seawater. Large amounts of water are absorbed into the rock. As the ocean floor subducts into the mantle, rising temperatures release this water again and it percolates upward. The water lowers the melting temperature of the surrounding rock, which partially melts to form new minerals. After several rounds of partial melting and reprocessing, granite forms, which is the basis of modern continents. Granite contains large amounts of silica and is less dense than other rocks. Continents made of granite float on top of the heavier rocks that make up the mantle, even when the granite is cold.

Continents formed in this way may have begun to take shape soon after Earth formed. They have continued to grow for much of the planet's history. There have been periods of rapid growth and relative quiet, including a major episode of continent formation around 2.5 billion years ago. Continental crust typically survives for much longer than oceanic crust, but it can be destroyed, especially at boundaries between plates. The formation and destruction of continental crust has slowed down over time as Earth cooled and the mantle began to convect less vigorously.

The gradual motion of the continents across Earth's surface is a central theme of modern geology. However, the theory of plate tectonics became widely accepted only relatively recently, in the 1960s. Before this, scientists such as Alfred Wegener had noted that some of the continents appear to fit together remarkably well, like pieces in a jigsaw puzzle. Rocks and fossils on the facing edges of different continents often have much in common, which also suggests they were once joined together and later moved apart. However, the notion of a static, rigid Earth was so ingrained that most geologists discounted these observations.

The most convincing piece of evidence appeared in the 1950s and 1960s with the advent of large surveys of the ocean floor. Geologists found that rocks beneath the ocean had different ages, with the youngest lying closest to midocean ridges. The rocks formed long, striped patterns running parallel to a ridge, each with a different age and magnetic properties, mirroring the pattern of stripes on the opposite side of the ridge. Clearly new ocean crust was forming at ridges and spreading apart, which meant that older crust was being destroyed elsewhere on the planet, at subduction zones. This discovery, together with the realization that most earthquakes occur in narrow zones where plates meet and collide, convinced scientists that plate tectonics is the main driving force shaping Earth's surface. For at least several billion years, the continents have repeatedly drifted with respect to one another, periodically coalescing into a giant supercontinent before parting ways again.

We don't know whether plate tectonics was firmly established in the Hadean, or whether it began a little later. Geologists usually learn about a period in Earth's history by examining rocks that formed at that time, but this approach won't work for the Hadean. The oldest known rocks on Earth formed around 4 billion years ago, when the Hadean was already drawing to a close. A combination of subduction, impacts, and weathering appears to have destroyed all rocks that formed earlier than this, so none survive intact today. Instead, much of what we know about early Earth comes from studying tiny rock fragments called zircons, which

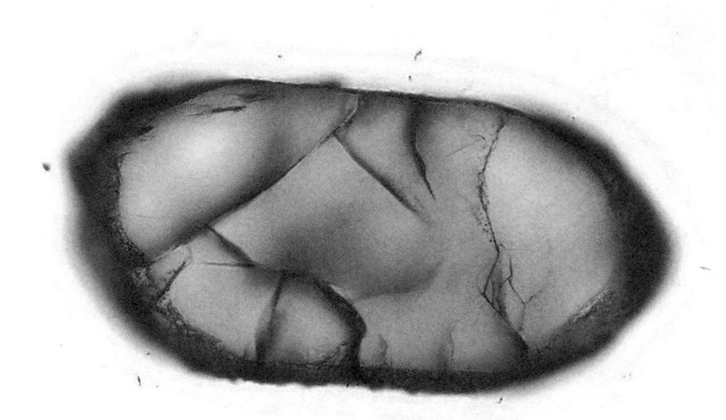

Figure 11.1. A Hadean zircon about 200 micrometers long, magnified 200 times and viewed in transmitted light. It was found in the Jack Hills in Western Australia at the Eranondoo Hill locality, where the world's oldest zircons have been documented. (Courtesy Stephen J. Mojzsis, University of Colorado)

we met in Chapter 4. These incredibly tough minerals are insoluble and strongly resist weathering and most disruptive geological processes. Because of these characteristics, zircons often preserve information about the conditions on Earth when they formed, even though their parent rocks have been thoroughly altered or destroyed.

The oldest known zircons, found in what is now Australia, formed about 4.4 billion years ago—not long after Earth itself (Figure 11.1). Judging by their composition, these minerals formed in continental crust out of magma that contained large amounts of water. The zircons also appear to have formed in a place where rocks were being weathered on the surface by liquid water to form sediments. In these ways at least, the Hadean Earth resembled the world of today.

The Hadean ended as it had begun, with a bang. Around 3.9 billion years ago, after millions of years of relative quiet, Earth, the Moon, and the other planets were pummeled by a series of large impacts—the late heavy bombardment that we encountered in the previous chapter. The largest collisions on Earth created impact basins 1,000 km (600 miles) or more in diameter, sending huge plumes of debris into space. Some of this debris traveled around the planet landing elsewhere on the surface, creating secondary craters. The tremendous heat released from these collisions boiled the upper layers of the oceans and baked the ground to high temperatures. If life existed on Earth at this point, it would have perished unless it was buried deep underground. At the same time, impacts may have lofted some rocks gently into space, moving just fast enough to escape Earth's gravity. These rocks became tiny asteroids orbiting the Sun. A few of them were destined to recollide with Earth many thousands of years later, landing as meteorites. Hardy microorganisms living inside these rocks may have survived this journey, reseeding our planet with life.

THE TREE OF LIFE

Nobody knows when life first appeared on Earth. There is substantial evidence that primitive life was firmly established at an early stage in the Archean, the period in Earth's history lasting from the end of the Hadean until about 2.5 billion years ago. The presence of ancient life is usually identified by means of fossils buried in sedimentary rocks that form when sand and clay particles settle gently to the ocean floor. Archean sedimentary rocks have survived in only a handful of places in the world, such as southern Africa, Greenland, and Australia. Tiny microfossils, fossilized microbial mats called stromatolites, or other signs of primitive life have been found at each of these sites, dating back to about 3.5 billion years ago. This means that life was already widespread by this time.

No fossils have been found that are older than this. However, there is one piece of circumstantial evidence to suggest that life existed at an even earlier stage. Carbon has two stable isotopes: carbon-12 and carbon-13. Since life on Earth prefers to use the lighter isotope, rocks formed from biological materials tend to be enriched in carbon-12 compared to other rocks. Some carbon-bearing materials preserved in 3.9-billion-year-old rocks in Greenland are enriched in exactly this way. Perhaps life was flourishing in at least one place on Earth 3.9 billion years ago, right after the late heavy bombardment ended.

All living things on Earth share certain characteristics. They are built from proteins composed of the same 20 amino acids. They store their genetic information using molecules of DNA and transmit this information using the related molecule, RNA. They make use of the same chemical pathways to reproduce, generate energy, and manufacture proteins. These similarities almost certainly mean that all living organisms are descended from the same ancestor that lived billions of years ago. Over time, millions of new species have emerged and evolved in different ways, but all have retained a common set of biological tools developed in the distant past.

Scientists have tried to map out how different organisms are related to one another using differences in their genetic codes. Species with a similar genetic makeup diverged from each other relatively recently, while species that are genetically very different split apart at a much earlier stage. By looking at genes from thousands of different species, scientists have put together a "phylogenetic tree," commonly called "the tree of life"—a branching diagram showing how all the main groups of organisms are related to one another (Figure 11.2).

The tree of life has three main branches. All living organisms and their ancestors fall into one of these three domains: bacteria, archaea (single-celled organisms superficially similar to bacteria but very dif-

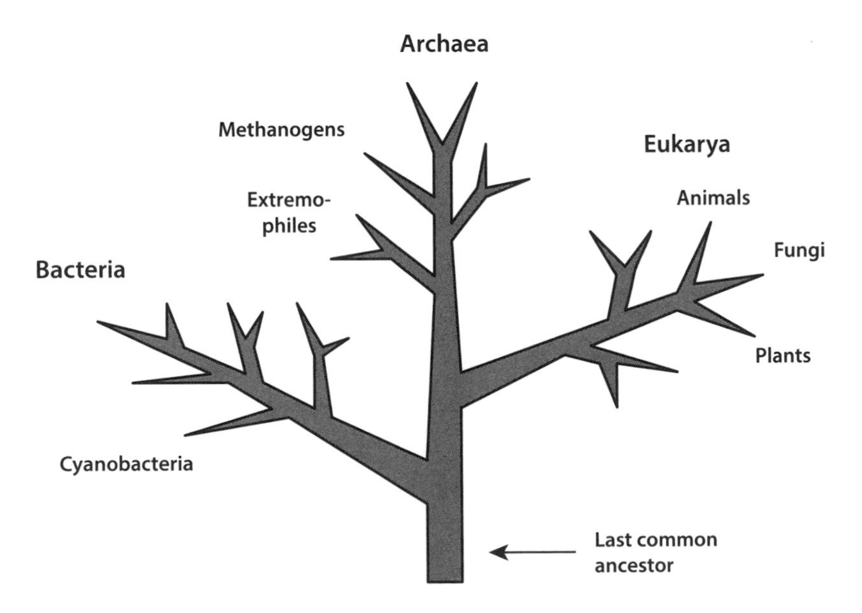

Figure 11.2. A simplified version of the "tree of life."

ferent genetically), and eukarya. The eukarya include all multicellular organisms such as trees, fungi, fish, and humans. This may seem like a very diverse group of life forms, yet it represents only a single branch on the tree of life, and the youngest one at that. At the base of the tree of life lies a shadowy creature called the "last common ancestor," a species that died out long ago yet whose progeny now fill every niche on the planet. We have no way of knowing whether the last common ancestor was the first organism that appeared on Earth, only that it was the first one to leave surviving descendants.

Many of the organisms near the base of the tree of life are "extremophiles," organisms that can withstand extreme temperatures or highly acidic or saline conditions that would kill other living creatures. Extremophiles thrive in hot, volcanic springs, and around hydrothermal vents on the ocean floor, living in superheated, pressurized fluids with temperatures above 100°C (212°F). Life in these extreme environments is harsh, as one might imagine, and it probably took evolving organisms a long time to adapt to such conditions. For this reason, it seems unlikely that the first organisms on Earth were extremophiles. However, the presence of extremophiles near the base of the tree of life suggests that conditions on Earth were once so inhospitable by our standards that only extremophiles could survive. Perhaps life arose before the late heavy bombardment, and extremophiles were the only organisms that survived this cataclysm, so that their descendants repopulated the planet after the bombardment ceased.

THE BUILDING BLOCKS OF LIFE

Living cells are made of a few key polymers—long, complex molecules constructed from simpler units such as amino acids and sugars. The polymers include nucleic acids, such as DNA and RNA that store and transfer the cell's genetic information, and proteins that act as catalysts to promote particular chemical reactions. These chemicals are surrounded by a cell membrane, made of fatty molecules called lipids. Membranes are partially porous, so nutrients and waste products can move in and out of cells in a controlled manner. All polymers used by life on Earth are made from the same handful of elements: carbon, hydrogen, oxygen, nitrogen, sulfur, and phosphorus.

In the 1950s, Stanley Miller and Harold Urey performed a series of experiments in an attempt to simulate conditions early in Earth's history. Their goal was nothing less than to form the basic building blocks of life in the laboratory. Miller and Urey set up a sealed glass vessel containing a mixture of gases that were thought to be similar to Earth's early atmosphere: hydrogen, ammonia, and methane. Warm water was cycled through the vessel again and again to mimic rain passing through the atmosphere. Electrodes stuck into the vessel produced sparks, which acted like miniature bolts of lightning. Miller and Urey left the experiment running for a week and then came back to see what had happened. What they found was that the simple gases had been transformed into an exotic soup of organic materials. These included several amino acids, the building blocks of proteins. More recent experiments that include hydrogen cyanide in the mix of gases have also made nucleobases, one of the key components of DNA.

Scientists now think that early Earth's atmosphere contained a mixture of gases different from that used by Miller and Urey. We now know that ammonia is easily broken apart by ultraviolet light from the Sun to form hydrogen, which escapes to space, and nitrogen, which stays behind in the atmosphere. In addition, it is likely that carbon dioxide was more common on early Earth than methane. Making organic molecules from nitrogen and carbon dioxide is harder than it is using ammonia and methane, but experiments show that amino acids and other biologically useful chemicals can still form in the presence of certain minerals such as iron pyrite, carbonates, and clays. All of these minerals would have been abundant in rocks on early Earth. Clays are particularly interesting since they form a platform on which simple organic molecules might be able to arrange themselves into polymers similar to RNA.

Making the complex organic molecules used by living organisms requires energy. There were many sources of energy on early Earth, but some would have been more useful than others. Ultraviolet light can form organic chemicals, but it also breaks them apart just as readily. Lightning flashes may have been more useful since they occur close to the ground, allowing molecules formed by lightning to reach lakes and oceans intact before ultraviolet light broke them apart again. Lightning strikes can also make hydrogen cyanide, a highly versatile building block from which many organic molecules can be constructed.

Underwater volcanic vents could have been a prime location for synthesizing organic molecules before life emerged. High temperatures near a vent allow rapid reactions to take place. Newly synthesized molecules soon flow away from a vent to where the surrounding seawater is cold enough to preserve these fragile materials. On land, complex organic molecules may have formed in small ponds or at the edges of lakes where the evaporation of water increased the concentration of useful chemicals. Some meteorites contain significant amounts of organic chemicals, including amino acids. Meteorites may have been an important source of these materials on early Earth since meteorites fell at a much higher rate in the past than today.

Under suitable conditions, lipid molecules spontaneously coalesce to form spherical membranes. Such membranes form a sheltered environment for chemical reactions and could have been the precursors of the first cells. Membranes are porous enough to allow small molecules to diffuse in from outside. These molecules can then combine to form larger polymers that become trapped inside. Once enough chemicals have accumulated inside a membrane, the pressure may cause it to burst, releasing the newly formed polymers into the environment. In this way, polymers from different membranes can mix together and become incorporated into new membranes.

A bewildering variety of organic polymers must have formed and broken apart again in the distant past. Most polymers didn't amount to much. Every once in a while though, a chance encounter formed a polymer that could act as a catalyst, helping to assemble smaller molecules to make exact copies of itself. Over time, these special polymers proliferated at the expense of others. A kind of natural selection took place, just as it does among living creatures today.

RNA is one of the molecules life uses to store and transmit genetic information, but it can also act as a catalyst. Some of the earliest forms of life may have depended solely on RNA molecules, a situation that has been called the "RNA world." Protein formation probably evolved later, starting with a handful of amino acids that were abundant in the environment, then continuing with the addition of others that were rarely produced naturally. Finally, life developed DNA, which provides a more stable and secure way to store information than RNA. All three branches of the tree of life use DNA, which suggests that it was adopted at a relatively early stage. Even RNA is probably too complicated to have formed spontaneously, so it's likely that the RNA world was preceded by an even earlier generation of catalytic molecules whose identity remains a mystery.

This sequence of events seems like a plausible account of how life on Earth evolved. We should note, however, that this scenario is based only on what we can glean from organisms that are alive today, together with laboratory experiments and a good deal of educated guesswork. We have no direct evidence of life's beginnings in the fossil record, and it is likely that none has survived. As a result, the story of how life began remains somewhat speculative for now.

THE RISE OF OXYGEN

Living organisms require a source of energy to build the chemicals necessary to survive and reproduce. Early life probably derived energy from a variety of chemical sources using materials that occurred naturally in the environment. Creatures near the base of the tree of life often obtain energy from reactions involving sulfur, which is abundant in volcanic regions. Others make use of iron-bearing minerals. An important group of organisms combines hydrogen from volcanic vents with carbon dioxide from the atmosphere to produce energy and also to acquire the carbon needed to build organic molecules. These methanogens release methane, a strong greenhouse gas that may have played an important role in the climate of early Earth, as we will see.

At some point, life discovered how to use energy from sunlight to power chemical reactions, the process we call photosynthesis. Early photosynthesizers were very different from those today. They combined carbon dioxide with hydrogen sulfide to form useful organic materials and store energy for later use. Their main waste product was sulfur, a relatively innocuous material.

Toward the end of the Archean, a class of organisms emerged that had a profound effect on every other creature on Earth. These were the blue-green algae, or cyanobacteria. They found a way to use sunlight to combine carbon dioxide with water, a much more powerful reaction than those used by earlier photosynthesizers, but there was a price to pay for this great leap forward. Instead of producing harmless sulfur as a waste product, cyanobacteria put into the atmosphere something that Earth had not seen there before: oxygen.

Today, we tend to think of oxygen as a vital, life-sustaining substance. However, oxygen is also a highly reactive and corrosive gas, and it has taken billions of years for life to adapt to it and to overcome its more dangerous aspects. Back in the Archean, oxygen was a deadly toxin for almost every creature that encountered it. As cyanobacteria multiplied and flourished, and oxygen began to accumulate in the environment,

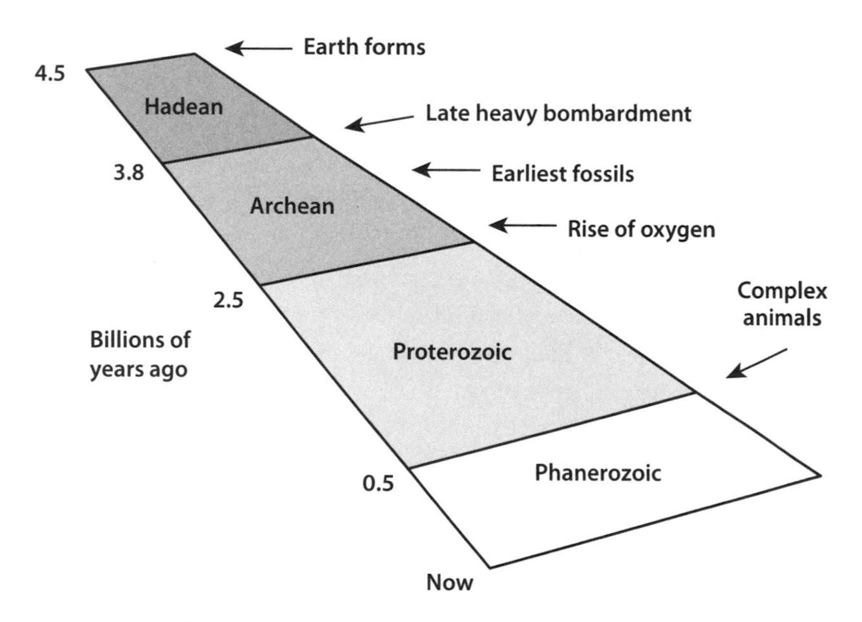

Figure 11.3. A timeline for the development of life on Earth, showing the four main eras since Earth formed. The numbers on the left are billions of years before the present.

every other organism had to find a way to adapt or perish. Some organisms eventually learned how to tolerate oxygen and even thrive in its presence. Others, such as the methanogens, were driven underground or deep underwater, to places were oxygen didn't penetrate.

Scientists can see the changing state of Earth's atmosphere reflected in the geological record. Perhaps the clearest examples are banded iron formations. These are sedimentary rocks that contain numerous alternating layers of red and gray material. The red layers are rich in iron oxides, while the gray layers contain little iron. Back in the Archean, Earth's oceans probably contained large amounts of iron dissolved in the water. As oxygen appeared in large quantities, the iron became insoluble and sank to the sea floor, forming a red layer atop earlier gray silts. It appears that the oxygen levels fluctuated substantially as life and the environment adjusted to the new state of affairs. Sediments laid down on the sea floor followed these trends, alternating between iron-rich and iron-poor deposits until the modern oxygen-rich atmosphere was firmly established.

The rise of oxygen led to major upheavals for life on Earth, but not all these changes were bad. As oxygen built up in the atmosphere, reactions involving sunlight produced the form of oxygen molecule that we call ozone. This highly reactive gas is poisonous at ground level, but high up in the atmosphere its presence is beneficial. Ozone has a remarkable ability to absorb ultraviolet light from the Sun, shielding life below from the damaging effects of this radiation.

Prior to the appearance of oxygen, the level of biological activity on Earth was low since the available chemical reactions tended to produce relatively little energy. Reactions between oxygen and organic materials generate large amounts of useful energy, so the rise of oxygen was accompanied by the appearance of a whole new class of energy-hungry creatures, the eukarya. These organisms have cells with complex internal structures, including a nucleus where the genetic information is stored. The eukarya represent the third branch on the tree of life. They include humans and all the advanced animals living on Earth today. From our point of view, the rise of oxygen was definitely a positive development.

A FAVORABLE CLIMATE

Earth during the Archean was very different from the world today, and yet temperatures seem to have been just right for life to exist. The temperature on Earth is set by a balance between incoming energy from sunlight and outgoing infrared radiation. The amount of solar energy reaching the ground is controlled by how effectively Earth reflects radiation back to space, a quantity called albedo. The amount of infrared radiation leaving Earth depends on the temperature and also the presence of greenhouse gases, such as carbon dioxide, in the atmosphere. Greenhouse gases absorb some infrared radiation, so Earth's surface has to become hotter in order to radiate away enough heat to balance the incoming energy from sunlight. Taking these factors into account, a simple calculation shows that Earth's average temperature today should be about 15°C (59°F), which is just what we observe.

The same calculation can be used to estimate temperatures early in Earth's history. One important difference is that the Sun has changed over time. In the past, less of the Sun's hydrogen fuel had been converted into helium, so the Sun was less dense than it is today. Nuclear reactions happened more slowly then, so the Sun was about 30 percent fainter when Earth first formed than it is now. Earth should have been colder as a result—so cold in fact that it would have been completely covered in ice. But we know from the zircons and the existence of ancient fossils that early Earth was warm enough to have liquid water. This discrepancy, called the "faint young Sun paradox," has perplexed scientists for decades.

The most likely solution to this conundrum is that Earth's atmosphere contained a lot more of greenhouse gases in the past than it does today. It isn't clear what those gases were, but carbon dioxide and methane are the most obvious candidates. Methane is particularly promising in this respect since it is only present in small amounts in Earth's atmosphere today. Today, methane is rapidly removed from the air by reactions with oxygen. However, oxygen wasn't present in the atmosphere early in Earth's history, so methane levels could have been substantially higher than today.

Earth's climate seems to have been remarkably stable over billions of years, with temperatures generally between the freezing and boiling points of water. In 1981, James Walker suggested that Earth's surface temperature is kept within this narrow range by a natural thermostatic process in which plate tectonics plays a crucial role. The thermostat works like this. Carbon dioxide in the atmosphere dissolves in rainwater to form carbonic acid. This acid gradually eats away exposed silicate rocks, and the products of this weathering are transported by rivers to the ocean. Here, the weathering products form carbonates that sink to the ocean floor and are buried in sediments. Millions of years later, when the ocean floor is subducted into Earth's mantle, the carbonates break down, releasing carbon dioxide back into the atmosphere through volcanoes. Scientists call this sequence of events the "carbon-silicon cycle" (Figure 11.4).

Walker's key insight was that the weathering of rocks requires liquid water, and weathering proceeds faster in warm water than in cold water. Carbon dioxide is an important greenhouse gas, so if it begins to build up in the atmosphere, the temperature will rise, weathering will happen more rapidly, and carbon dioxide levels will fall again as a result. If carbon dioxide becomes scarce, weathering slows down, and carbon dioxide escaping from volcanoes will begin to build up in the atmosphere. In both cases, the carbon-silicon cycle works as negative feedback, helping to stabilize the climate and keep carbon dioxide levels and the temperature at a happy medium.

Life is not an essential ingredient of the carbon-silicon cycle, but it plays an important role. Microorganisms in soil help to speed up weathering by scouring nutrients from rocks. The appearance of plants on land half a billion years ago also helped to increase the weathering rate. In the ocean, many creatures use weathering products to make carbonate shells, which sink to the ocean floor when their owners die, thus removing carbon from the system.

James Kasting and his colleagues have extended Walker's idea to show that any Earth-like planet should have liquid-water temperatures at the surface due to the carbon-silicon cycle provided that the planet lies within a particular range of distances from its star. This range is commonly called the star's habitable zone. Its precise extent depends on

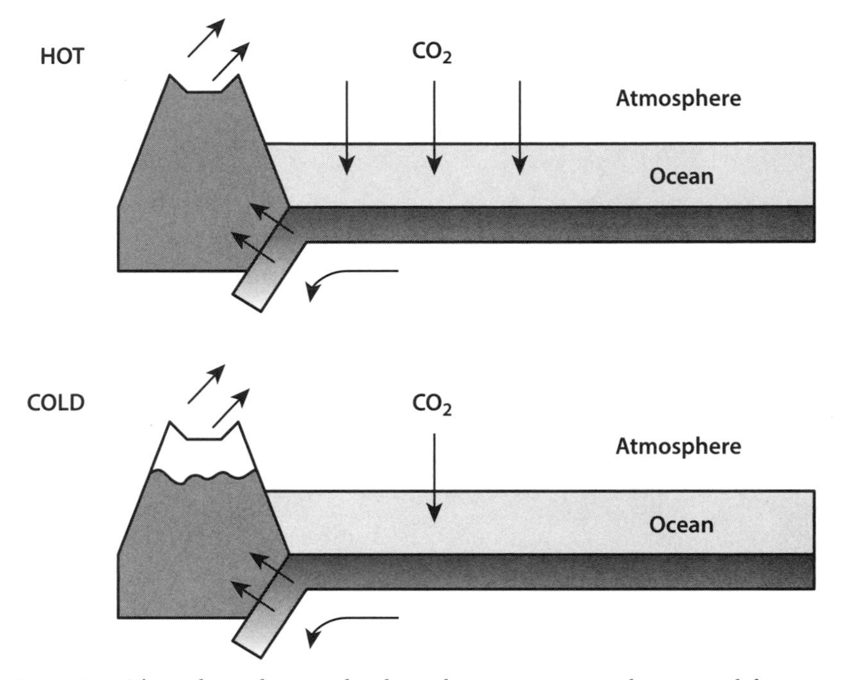

Figure 11.4. The carbon-silicon cycle—how plate tectonics contributes to stabilizing Earth's climate. If carbon dioxide builds up in the atmosphere and the climate heats up (*top*), rocks weather more quickly. As a result of the weathering process, carbon dioxide is captured into solid carbonates that sink to the ocean floor. Atmospheric carbon dioxide levels then decline and the climate becomes cooler (*bottom*). Later, when the sedimentary rocks containing carbonates are subducted into Earth's mantle, the carbonates break down and carbon dioxide is released back into the atmosphere by volcanoes. The cycle then starts again.

how bright the star is. Earth lies comfortably within the Sun's habitable zone while Venus does not.

Surprisingly, Mars probably lies within the Sun's habitable zone, yet Mars has no liquid water on its surface and the average temperature is far below freezing. Mars's failure to be "habitable" has more to do with its small size than its location. Mars appears to be too small to undergo plate tectonics or crustal recycling of any kind. With a diameter only half that of Earth's, Mars has cooled more rapidly than our own planet, and today it is largely a geologically dead world. Mars's weak gravity means that most of its atmosphere has escaped into space, a process made worse by the lack of a magnetic field to protect the atmosphere from erosion by energetic particles coming from the Sun. With no crustal recycling, there is no way to replenish these lost gases, and the residual atmosphere is too thin to provide much of a greenhouse effect. Early in its history, Mars froze. It has remained frozen ever since.

SNOWBALL EARTH

Despite the stabilizing effects of the carbon-silicon cycle, Earth may have been completely covered in ice several times in its history. When ice forms in large amounts, it produces glaciers that slowly travel downhill due to their immense weight. As a glacier grinds downhill, it leaves behind telltale grooves or "striations" in the underlying bedrock. At the same time, pebbles and boulders called "dropstones" are carved out of the bedrock and carried along for substantial distances before being deposited in sediments at the end of the glacier. By searching for striations and dropstones in ancient rocks, geologists can map out regions that were covered in ice in the distant past.

Some minerals are magnetic, and when these form sedimentary rocks, they tend to orient themselves so that they are aligned with Earth's magnetic field. By measuring a rock's magnetization, geologists can deduce the latitude at which it formed. If a magnetized rock layer contains dropstones and striations, it tells us that there were glaciers at this latitude. Strikingly, at several times in the past, glaciers appear to have covered much of Earth's surface, perhaps extending from the poles all the way to the equator. Scientists call these bizarre occurrences "snowball Earth" episodes.

It turns out that Earth's climate actually has two stable modes: one where the surface is almost ice-free, as it is today, and one in which Earth is almost completely covered in ice. Earth can exist in either mode for the same level of sunlight and the same amount of greenhouse gas because the planet's albedo is very different in each case. The transition to a snowball Earth can be rapid. Suppose that a minor variation in the climate causes more snow to fall and accumulate at high latitudes. Snow and ice reflect away most of the sunlight that falls on them. Consequently, Earth absorbs less energy and cools, leading to more snow accumulation, and so on. This positive feedback loop can quickly transform Earth's climate from its current balmy state to snowball Earth.

One of the earliest snowball Earth episodes apparently coincided with the rise of oxygen in the atmosphere around 2.3 billion years ago. This makes sense if the atmosphere on early Earth contained large amounts of methane. Oxygen reacts rapidly with any methane in the air, converting it to carbon dioxide. Both methane and carbon dioxide are greenhouse gases, but methane is by far the more powerful of the two. When oxygen first appeared, methane levels would have dropped precipitously, reducing the greenhouse effect and launching a global ice age.

The snowball Earth episodes were quite short-lived, lasting for less than 10 million years. How then did Earth manage to escape from its snow-covered state? An important clue comes from the geological record. The rocks laid down immediately after Earth recovered from its glaciations typically contain thick deposits of limestone and similar carbonate-bearing rocks. These "cap carbonate" rocks apparently formed under unusual conditions when Earth's atmosphere contained large amounts of carbon dioxide. This makes sense. When Earth's surface was covered in ice, weathering of rocks ceased. Carbon dioxide continued to be released from volcanoes, seeping through cracks in the ice and building up in the atmosphere. Eventually, carbon dioxide levels became so high and the greenhouse effect so strong that the glaciers melted. Huge areas of freshly scoured rock were suddenly exposed to large amounts of carbon dioxide, leading to a burst of weathering and the formation of the thick cap carbonate layers.

Snowball Earth episodes would have been tough for life, but some organisms found a way to survive. Even if Earth's oceans were completely covered in ice, key biological processes such as photosynthesis would have continued in the water beneath the ice as long as the ice was no more than a few meters (several feet) thick. It is also possible that the ice sheets didn't reach all the way to the equator, leaving a narrow band of open water or slushy ice. Volcanically active regions may also have provided small oases where life continued.

FUTURE HABITABILITY

The Sun will continue to grow brighter in the future as it converts more of its hydrogen into helium. This will pose more problems for life on Earth, perhaps the sternest test it has faced yet. Within the next billion years or so, the Sun will be so bright that no greenhouse gases will be needed to keep Earth's surface at liquid-water temperatures. At this point, weathering of rocks will remove almost all carbon dioxide from the atmosphere, making it impossible for modern plants to grow. More seriously still, the carbon-silicon cycle will no longer be able to regulate temperatures on Earth, and it will be only a matter of time before our planet shares the same fate as Venus.

If intelligent life exists at this stage, it could bring advanced technology to bear on the problem. One approach would be to change Earth's orbit by gradually moving Earth away from the Sun as the Sun grows hotter. This is not as far-fetched as it sounds. Scientists recently worked out that Earth's orbit could be changed by diverting a large asteroid so that it repeatedly swings past Earth and its gravitational pull gradually alters Earth's course at each encounter. If the same asteroid also repeatedly swings past Jupiter, the asteroid's own orbit is similarly affected. The net effect is to gradually lift Earth away from the Sun using energy taken from Jupiter's orbit.

Such high-tech solutions are clearly not available to more primitive forms of life, but they could find equally effective ways to adapt. Life might inject into the atmosphere particles that form aerosols or hazes, enabling Earth to reflect away much of the incoming sunlight and keep cool even as the Sun grows brighter. Life has endured numerous crises during Earth's history and has always found a way to adapt. There is every hope that, with life's help, Earth will continue to be habitable for a long while yet.

TWELVE

WORLDS OF GAS AND ICE

GIANTS OF THE SOLAR SYSTEM

Imagine if we could take a voyage deep into the interior of Jupiter, the most massive planet in the solar system. What would we find? As we first approach the planet, we would see its familiar exterior: colorful belts of clouds encircling the planet, some white, others various shades of red, orange, and brown (Figure 12.1). Numerous oval features are dotted at intervals between the cloud belts, including the famous great red spot.

As we enter Jupiter's atmosphere, this flat image takes on a three-dimensional structure. Belts of various colors resolve themselves into cloud banks at different altitudes. The spots become giant rotating storms that rise above the surrounding layers. Descending through the clouds, we find layers of ammonia, ammonium sulfide, and water. Deeper still lie clouds containing rocky and metallic elements. Sampling the atmosphere as a whole, we find it is mostly composed of hydrogen and helium with a smattering of other gases such as water and methane.

The temperature and pressure rise steadily with depth until matter becomes compressed so tightly that it behaves more like a liquid than a gas. Heat welling up from deep within the planet drives convective eddies—huge plumes of fluid rising and falling, continually mixing the planet's interior. Droplets of helium and neon form a continual drizzle descending through the maelstrom. Deep within Jupiter's fluid interior, the pressure becomes so great that electrons are stripped away from the hydrogen atoms to form a material that behaves like liquid metal. Finally, near the very center of the planet, we arrive at a small, dense core

Figure 12.1. Jupiter, imaged by the Hubble Space Telescope in 2009. The temporary dark spot toward the lower edge of the disk, just right of center, was created by an impact. (NASA, ESA, H. Hammel [Space Science Institute, Boulder, Colorado], and the Jupiter Impact Team)

composed of heavy elements compressed to an almost unimaginable degree.

Such a voyage is likely to remain hypothetical for the foreseeable future, but it is not entirely speculative. In 1995, the Galileo spacecraft launched a probe that embarked on the first leg of this journey. The probe entered Jupiter's upper atmosphere, deployed a parachute, and gradually descended, returning data on the atmosphere's composition, temperature, pressure, and cloud layers for almost an hour. The atmosphere proved to be mostly composed of hydrogen and helium, as expected. Surprisingly, it was also very dry, apparently because the probe entered a localized weather system where the atmosphere had lost most of its water. Other gases were present in larger amounts, including methane, ammonia, hydrogen sulfide, and heavy noble gases. Eventually the temperature and pressure of Jupiter's atmosphere became too much to withstand, and the Galileo probe fell silent, but not before it had given us a fascinating peek inside the largest planet.

It's also possible to deduce some things about Jupiter's interior by observing it from afar. The planet's low density tells us that it must be mostly composed of the lightest elements, hydrogen and helium. Measurements of Jupiter's gravitational field show that it is centrally condensed and probably has a dense core containing around 5 to 10 Earth masses of material. The planet's powerful magnetic field means that much of the interior is electrically conducting. Here on Earth, laboratory experiments that simulate the tremendous pressure inside Jupiter suggest that hydrogen behaves like a metal under such conditions. Astronomical measurements also tell us that Jupiter releases more heat than it receives from the Sun as material is gradually compressed by gravity. Jupiter is slowly shrinking over time and must have been substantially larger when it first formed.

We know somewhat less about the other giant planets of the solar system since no probes have gone down through their atmospheres and examined them directly. However, remote observations show that they have much in common with Jupiter. The low densities of all four giants mean they are mostly made of much lighter stuff than their terrestrial cousins. As on Jupiter, most of this bulk is gaseous in the outer layers but must be compressed into liquids in the interior. None of the giants has a solid surface, and the transition between gas and liquid is not a sharp one. Astronomers refer to the outer part of the fluid envelope as the atmosphere, although the depth of the base of the atmosphere is rather arbitrary. Because the giant planets are all so far from the Sun, their upper atmospheres are frigidly cold, ranging in temperature from about −160°C (−256°F) for Jupiter down to about −220°C (−364°F) at Uranus and Neptune.

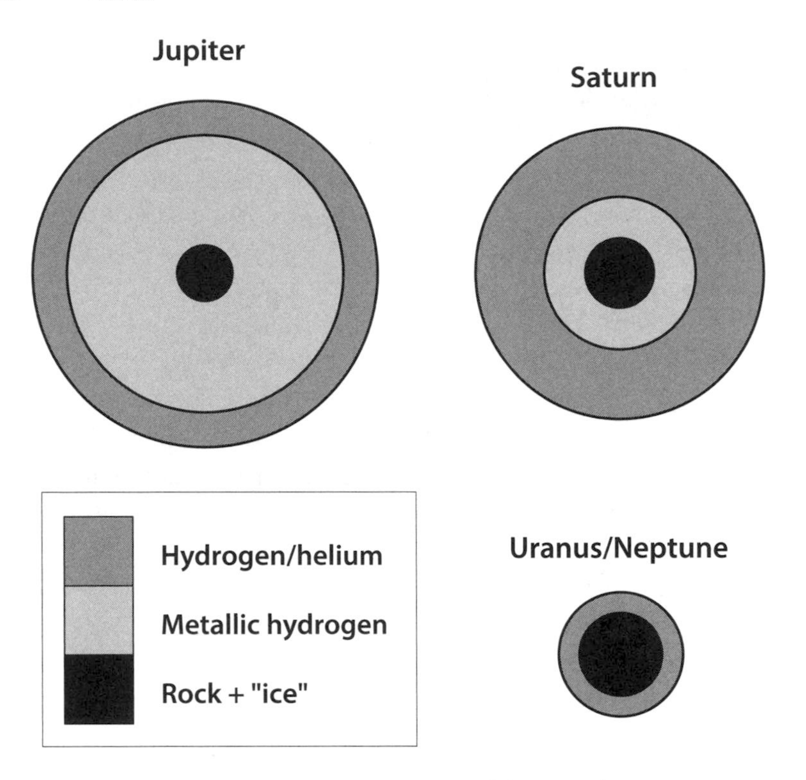

Figure 12.2. The interior structure of the giant planets.

The gases that make up the giant planets' atmospheres are largely transparent, but small amounts of methane in the atmospheres of Uranus and Neptune absorb red light strongly, giving these planets their characteristic bluish colors. All four giants have clouds made of ice crystals or liquid droplets, and hazes of fine particles condense at various levels within their atmospheres, just as they do on Earth. The giants' rapid rotation tends to streamline these clouds into narrow bands of different colors, creating the planets' visible faces.

Despite these similarities, there are also substantial differences between the giant planets. Jupiter and Saturn are the real giants, similar in size to each other with diameters around 20 times that of Earth. Jupiter weighs more than 300 times Earth, and has over 3 times more mass than Saturn compressed into its globe, so the densities of the two planets are significantly different. Saturn is actually less dense than water. The

Table 12.1. Mass, diameter, and density of the giant planets

Planet	*Mass (Earth = 1)*	*Diameter (Earth = 1)*	*Average density (water = 1)*
Jupiter	318	22.4	1.33
Saturn	95	18.9	0.69
Uranus	15	8.0	1.27
Neptune	17	7.8	1.64

other giants, Uranus and Neptune, are smaller and less massive (Figure 12.3). Both are around 8 times the size of Earth and have masses 14 and 17 times Earth's, respectively. Packing in more mass, Neptune is somewhat denser than Uranus. Unlike Jupiter and Saturn, the smaller giants contain only about 10 to 20 percent hydrogen and helium, and they are mostly composed of heavier elements.

Given these differences, astronomers have subdivided the giant planets into two categories. Unfortunately, the terminology is rather misleading. Jupiter and Saturn are called gas giants because they are mostly

Figure 12.3. Uranus (*left*) and Neptune (*right*). The image of Uranus was taken by the Hubble Space Telescope (HST) in 2003. Filters were used to show up cloud features, and the brightness of the area outside Uranus was enhanced to reveal the ring system and several moons. (NASA and Erich Karkoschka, University of Arizona) The image of Neptune was taken by Voyager 2 on August 14, 1989. It shows the Great Dark Spot, which had disappeared by the time HST images were taken in 1994, though another similar spot emerged subsequently in a different place. (NASA/JPL)

made of hydrogen and helium, which we normally think of as gases. However, at the tremendous pressures inside Jupiter and Saturn, these "gases" behave more like liquids, and as we have seen, hydrogen becomes distinctly metallic in nature. Uranus and Neptune are described as ice giants because they are made mostly of materials that become ices when they are very cold. However, as with the gas giants, the high temperatures and pressures inside Uranus and Neptune mean that these "icy" substances are actually hot fluids.

Clearly, both kinds of giant planet are very different from Earth and the other terrestrial planets. How could objects with such disparate compositions form in the same solar system? In one respect we shouldn't be too surprised that there are planets mostly made of hydrogen and helium since these two gases dominated the solar nebula. However, holding on to hydrogen gas requires a powerful gravitational field. The giant planets have enough mass to do so, while the terrestrial planets do not. Thus, the compositions of the giant planets are intimately linked to their size and mass.

There are two schools of thought about how the giant planets formed. The more likely possibility, called the core accretion model, is that the giants began as solid bodies, larger analogues of the terrestrial planets. These objects became massive enough to capture and hold on to hydrogen and helium gas from the solar nebula. However, they must have done so very quickly. Hydrogen and helium don't condense into solids or liquids at the temperatures and pressures that existed in the solar nebula, which means that the giant planets must have acquired these elements in the form of gases. As we saw in Chapter 8, the gas-rich protoplanetary disks that surround young stars disappear after only a few million years. There is every reason to think that the solar nebula was similarly short-lived, so Jupiter and Saturn must have acquired most of their bulk in a short space of time.

The relatively short time available for Jupiter and Saturn to grow so large—too little time according to some astronomers—led planetary scientist Alan Boss to devise an alternative theory, known as the disk instability model. In this scenario, the gas-rich planets formed directly when portions of the nebula became dense enough for their own gravity to cause them to collapse and detach from the surrounding material. We now look at each of these scenarios in more detail.

BUILDING GIANTS BY CORE ACCRETION

Why would solid planets grow larger in the outer solar system than close to the Sun? There are two reasons. First, in the cool outer regions of the solar nebula, temperatures were low enough for ices to form. These ices include ordinary water ice together with frozen ammonia and carbon dioxide and, with somewhat lower freezing points, methane and carbon monoxide. The presence of ices meant that there was roughly twice as much solid mass available to build planets in the outer nebula as there was in the region where the rocky planets formed.

The second reason is more subtle. In Chapter 9, we saw how growing planetary embryos tend to sweep up material from a feeding zone—the region around their orbit where their own gravity is more important than the Sun's. In the outer solar system, an embryo's gravitational reach is greater than it would be nearer the Sun because the Sun's gravitational pull is weaker.

The combined effect of larger feeding zones and extra solid material in the form of ice allowed planetary embryos near Jupiter's orbit to grow substantially larger than those near Earth during the rapid oligarchic growth stage of planet formation. When oligarchic growth ended, the largest bodies in the inner nebula were probably about the size of the Moon or Mars, whereas planetary embryos might have grown to 10 Earth masses or more in the region where Jupiter and Saturn are today.

Once planetary embryos reached the size of Mars, their gravitational pull would have been enough to compress the nebular gas around them. In effect, the gas formed diffuse, extended atmospheres surrounding the embryos. These atmospheres were only temporary—if the nebula had suddenly vanished, the atmospheres would have disappeared too. However, while an atmosphere remained, it played an important role in the growth of a young planet. Many passing planetesimals that narrowly missed an embryo's solid surface instead traveled through its atmosphere. These planetesimals were slowed down by gas drag and became captured by the embryo's gravity, ultimately colliding with the surface. Thus, an embryo with an atmosphere grew a good deal faster than it otherwise would have done.

The amount of gas in an atmosphere was controlled by a balance between the inward pull of the embryo's gravity and outward pressure within the gas. As an embryo grew larger, its gravitational pull increased and the atmosphere became denser and heavier. However, this balancing act could not continue forever. A simple calculation shows that once an embryo exceeds a critical mass, its gravity is too strong for pressure forces to resist. When this point was reached, gas began to flow onto the planet from the nebula. The embryo became a solid core at the center of a gaseous envelope that grew inexorably until the supply of gas was shut off.

How fast a planet's envelope grew depended in part on its ability to radiate away heat. Heat escaping as infrared radiation allowed the envelope to cool and contract so that more material flowed inward. However, in-falling material had the opposite effect. It released additional energy that warmed the envelope, causing it to expand. This slowed down the accretion of gas. The presence of dust grains in the envelope made the gas less transparent, which reduced the amount of infrared radiation that could escape. Some dust grains were entrained with the inflowing gas and remained suspended in the envelope for some time. Additional dust grains were formed when captured planetesimals plunged through the envelope and were torn apart by aerodynamic forces from the surrounding gas. Eventually, dust grains coalesced into larger particles that sank deep into the interior, but these were soon replaced by newly arriving grains.

The dusty nature of planetary envelopes and the difficulty of radiating away heat imply that gas would have flowed inward slowly at first. Computer simulations suggest that it could take a million years for an envelope to become as massive as the solid core at the center of the planet. At this point though, growth sped up dramatically. The planet would have begun to undergo runaway gas accretion and could have grown to Jupiter's mass in as little as 10,000 years.

This naturally raises the question of why Jupiter stopped growing when it did. Several factors can slow or truncate the flow of gas onto a giant planet. First, a planet can accrete gas only as fast as the surrounding nebula can supply it. Once a planet has used up the gas in its vicinity, it takes time for gas to arrive from elsewhere in the nebula to replace it. Second, even if gas within the nebula is highly mobile, a planet's gravity can frustrate its own growth. Planets the size of Jupiter are massive enough for their gravity to strongly affect material in the surrounding region. The combined gravity of Jupiter and the Sun would have cleared a ring-shaped gap surrounding the planet's orbit, rather like the gaps opened by moons embedded within Saturn's rings. When a gap opened up in the solar nebula, the flow of gas onto the planet in the gap was interrupted. Finally, the solar nebula had a limited lifetime of only a few million years. Once the nebula had dissipated, the giant planets were unable to grow any larger.

The core accretion model naturally explains many of the features of the gas giant planets Jupiter and Saturn. The ice giants are a little harder to understand. Uranus and Neptune possess relatively small envelopes of hydrogen and helium amounting to a few Earth masses each. Presumably, these planets formed near the end of the life span of the nebula, and gas accretion had only just begun when the nebula disappeared. The timing here is not as critical as one might imagine since the early stages of gas accretion happen slowly and a planet can remain the size of Uranus or Neptune for quite some time.

A more problematic question is how solid cores grew to be the size of those within Uranus and Neptune in the first place. Growth rates declined dramatically with increasing distance from the Sun both because the solar nebula was more tenuous in the outer reaches of the solar system and because objects moved more slowly with respect to one another, so collisions were less frequent. Jupiter's core probably took a few million years at most to form, but a similar-sized body could have taken a billion years to grow at Neptune's current location. This is far longer than the lifetime of the solar nebula, which means that all the gas would have vanished long before Neptune grew large enough to acquire a gaseous envelope. The situation is less severe for Uranus, but it is still hard to see how it could have formed before the nebula dissipated.

For this reason, most scientists believe that the ice giants formed nearer to the Sun than they are today. All four giant planets could have been packed more closely together when they formed, provided that another process changed their orbits at a later stage. This process probably involved gravitational interactions with leftover planetesimals, as we will see in Chapter 14. Planet formation is an inefficient process, and there may have been tens of Earth masses of residual small bodies that didn't end up in one of the giant planets. Whenever one of these planetesimals passed close to a giant planet, mutual gravitational interactions would have slung the smaller body onto a substantially different orbit while at the at the same time nudging the planet's orbit in the opposite direction. Many of these planetesimals ended up in the Oort cloud or were ejected from the solar system entirely.

Computer simulations show that, as a result of these interactions, Jupiter probably ended up being a few tenths of an AU closer to the Sun than where it started. The orbits of Saturn, Uranus, and Neptune all moved outward, possibly by as much as 15 AU in the case of Neptune. For the most part, this transformation would have been gradual. However, there may have been periods of rapid change, causing dramatic upheavals in the rest of the solar system.

It isn't entirely clear why Saturn stopped growing before it became as large as Jupiter. Runaway gas accretion should have continued very rapidly for a planet of Saturn's mass. Saturn may have started to open a gap in the solar nebula, but the planet's gravity is only strong enough to partially clear the gas away from its vicinity. Some gas would still have made it across the gap and flowed onto Saturn if it was the only giant present. However, it is possible that Jupiter was already fully formed at this stage. Computer simulations suggest that the combined gravity of Jupiter and Saturn could have been enough to clear a complete gap surrounding both planets, especially if the two planets lay closer together than they do today. Saturn's growth would have halted prematurely—a permanent reminder of the presence of its larger sibling.

THE DISK INSTABILITY MODEL

The core accretion model for giant planet formation can explain a good deal and is favored by many scientists, but the disk instability model is a plausible alternative. The solar nebula was massive, even though it was mostly composed of tenuous gas. At a minimum, the nebula contained at least 1 percent of a solar mass (some 3,000 Earth masses), and it may have been substantially more massive. The combined gravitational influence of all this gas was considerable. In its cool, outer regions, the nebula may occasionally have become unstable as a result. Under these circumstances, gravity would have become strong enough to overcome motions within the gas, causing the nebula to fragment into clumps.

This is the same gravitational instability mechanism that we met in relation to planetesimal formation in Chapter 8, but in this case it affects the gaseous component of the nebula rather than solid particles. Calculations show that these instabilities take place on a vast scale, with a typical clump containing about as much mass as Jupiter. Clumps can form rapidly, in just a few orbital periods or a matter of decades. Once a clump forms, however, it is unclear what happens next. Some computer simulations show that clumps are rapidly broken apart by the rotation of the nebula. Other calculations find that the clumps remain intact and begin to shrink, destined ultimately to form giant, gas-rich planets.

The principal difference between these apparently contradictory calculations is how fast the clumps are assumed to cool. Rapid cooling leads to dense, stable clumps. Slow cooling results in clumps that break apart. Unfortunately, we don't know which assumption is correct at this point, so the disk instability scenario remains in a state of limbo—plausible but unconfirmed.

If the giant planets did form by disk instability, they would have begun life as homogeneous bodies with a composition identical to the solar nebula. Interestingly, such objects could have formed solid cores after they acquired their gas, exactly reversing the sequence of events in the core accretion scenario. Dust grains and planetesimal fragments swept along with the inflowing gas would have coagulated into larger clumps and settled toward the center of the clump while it was shrinking. This solid material would ultimately have coalesced into a dense core at the planet's center.

SPIN AND TILT

Whichever mechanism was responsible for the growth of the four giants, it produced planets that spin rather faster than their terrestrial counterparts (Table 12.2). This rapid rotation makes the giant planets bulge outward around the equator. The effect is particularly noticeable for Saturn, which is 10 percent wider across its equator than the distance between its poles. Gas in the atmosphere of each planet travels at

Table 12.2. Obliquity and rotation periods of the giant planets

Planet	*Obliquity (degrees)*	*Rotation period (hours)*
Jupiter	3	9.9
Saturn	27	10.7
Uranus	98	17.2
Neptune	30	16.1

immense speeds due to the rapid rotation, making it difficult for material to deviate to the north or south before the planet's rotation pulls it back again. For this reason, the various cloud formations are channeled into narrow bands, each occupying a different latitude.

While the giant planets were forming, the inflow of gas from the solar nebula probably ensured that each planet's spin axis was aligned more or less vertically with respect to its orbit. Today, the spin axes of Saturn and Neptune are tilted by about 30 degrees (Table 12.2), and Uranus is tilted over so much that it seems to lie on its side. Something has clearly changed the obliquity of these planets since they formed. In Chapter 9, we saw how gravitational and tidal interactions changed the obliquities of the terrestrial planets over time. Similar processes could have operated in the outer solar system too, but it seems unlikely that this was the whole story, especially for Uranus.

Collisions can also change the way a planet spins. A plausible explanation for Uranus's strange orientation is that it was struck at a glancing angle by a body with a mass several times that of Earth near the end of its formation. However, there is a problem with this hypothesis. In such a giant collision, Uranus's moons should have remained more or less on their original orbits. Today, Uranus's satellites travel in the same plane as the planet's equator, meaning the satellite system is also tilted on its side. How could this have happened? Recent computer simulations suggest a possible explanation.

As we will see, the satellites of the giant planets probably formed from a disk of material in orbit around each planet. If Uranus were tilted by a giant impact while this disk was still present, collisions between particles in the disk would soon realign the disk with the planet's equator again. Moons that later formed in this disk would naturally have orbits tilted by the same amount as Uranus. Unexpectedly, computer simulations show that these moons are likely to have retrograde orbits, traveling backward around the planet. This is not what we see today. There is, however, a scenario that can reproduce the actual Uranian system. It turns out that moons orbiting in the correct direction are the most likely outcome if Uranus suffered at least two large impacts rather than one. We cannot be sure whether this theory is correct, but if it is, large collisions may have played an important role in the growth of the outer planets as they did for Earth.

MASTERS OF MANY MOONS

One of the many differences between the planets in the inner and outer solar system is that the giant planets have extensive systems of satellites, rather like miniature planetary systems in their own right.

Jupiter's gravity has made it the master of at least 63 satellites, including the four Galilean moons—Io, Europa, Ganymede, and Callisto—which are the size of terrestrial planets (Figure 12.4). The Voyager and Galileo spacecraft found that these moons display a clear compositional trend with increasing distance from Jupiter. Io is composed entirely of rocky materials. Europa has a relatively thin veneer of ice and liquid water atop a large rocky interior. Ganymede and Callisto each contain substantial amounts of rock and ice. However, Ganymede is clearly differentiated, with a rock-rich interior and an ice-rich mantle. Callisto, on the other hand, is a more intimate mixture of ice and rock. Callisto's surface also appears to be very ancient compared to those of the other Galilean satellites. These observations place important constraints on how Jupiter's moons formed.

Together with four much smaller satellites lying nearer to Jupiter, the Galilean moons follow almost circular orbits close to the plane of Jupiter's equator and travel in the same direction as Jupiter's rotation. Astronomers call these "regular satellites" on account of their regimented orbits. The rest of Jupiter's retinue lies much farther from the planet, in orbits that are considerably less orderly. These "irregular satellites" all have very elongated, highly inclined orbits, and in most cases they are retrograde,

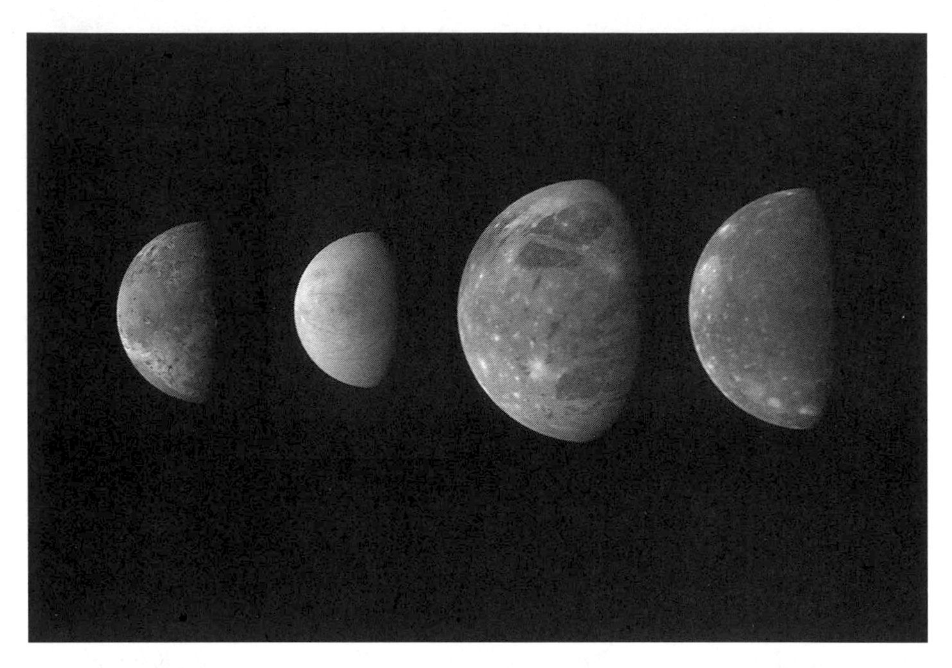

Figure 12.4. Jupiter's Galilean moons to scale, imaged by the New Horizons spacecraft. *From left to right*: Io, Europa, Ganymede, and Callisto. (NASA/Johns Hopkins University Applied Physics Laboratory/Southwest Research Institute)

revolving in the opposite direction to the planet's rotation. Typically, the irregular satellites are only a handful of kilometers (a few miles) across, and only two have diameters of 100 km (60 miles) or more.

Saturn's family of moons is quite similar in character to that of Jupiter, and the confirmed number, 62, is almost the same. However, Saturn has only one large satellite—Titan—comparable in size to the Galilean moons, together with half a dozen in the 400 to 1,500 km (250 to 1,000 miles) size range. Of the rest, 38 are irregular outer moons. Saturn also has a dozen small, inner satellites, and numerous tiny moonlets and clumps of material embedded within the planet's extensive system of rings.

The satellite family of Uranus is broadly similar to those of Jupiter and Saturn, but on a smaller scale. Neptune's is a little different. Both planets host regular and irregular moons. None of Uranus's 27 satellites are in the top league by size, but Triton, the largest of Neptune's 13 moons, ranks seventh in the solar system. It is unique among the large satellites because it is irregular, moving on a tilted, retrograde orbit.

FORMATION OF REGULAR SATELLITES

Computer simulations suggest that each of the giant planets would have been surrounded by a small disk of gas and dust while the planets were pulling in gas from the solar nebula. Gas entered this circumplanetary disk from the nebula, moved through the disk, and then fell onto the planet. It seems plausible that the regular satellites, such as the Galilean moons, formed out of solid particles moving within the circumplanetary disk, in a scaled down version of planet formation in the solar nebula. However, there was one important difference: the orbital periods of regular satellites are much shorter than those of the planets.

If all of the mass contained in the Galilean satellites were originally present as small particles in Jupiter's circumplanetary disk, it would require only about 1,000 years for fully formed satellites to appear. Such rapid growth would generate tremendous amounts of heat, more than enough to melt Callisto and allow denser materials to sink to its interior. Since Callisto is not differentiated, it must have formed far more slowly, taking at least 100,000 years to reach its current size. In addition, a massive circumplanetary disk would have been too hot to allow ice to form where Europa and Ganymede are today. Tidal interactions between growing satellites and a massive disk could also have forced these objects to migrate into Jupiter and be destroyed.

Instead, it seems likely that Jupiter's circumplanetary disk was quite tenuous, at least while the satellites were forming. A good deal of gas passed through the disk, but only a fraction of the total was present at any one time. Solid particles entrained with the incoming gas quickly coagulated into larger bodies that remained in the disk while the gas continued to flow inward onto the planet. Thus, the ratio of solid material to gas increased over time. The slow accumulation of solids within this gas-starved disk, and the disk's tenuous nature, meant that the moons took a long time to form and temperatures were low enough for ice to be present.

The regular satellites we see today probably appeared near the very end of their planet's formation. At this point, the inward flow of gas would have been small, the circumplanetary disk at its most tenuous, and tidal interactions too weak to cause moons to migrate into the planet. Most of the radioactive isotopes that were present early in the solar system would have decayed, and radioactive heating on Callisto would have been minimal. It is possible that earlier generations of satellites formed and fell into their planet and the ones we see today are the sole survivors. The masses of the satellite systems of Jupiter, Saturn, and Uranus are each about 0.01 percent that of their planet, so a self-regulating process could have been at work. It may be that, whenever moons grew larger than this, they migrated into the planet and the process of building satellites had to begin again.

The orbits of many regular satellites are in resonance with one another. The best-known example is the Laplace resonance, in which the orbital periods of Io, Europa, and Ganymede are locked in the ratio 1:2:4. Six pairs of satellites in the Saturnian system also lie in resonances. Resonances have quite a small probability of occurring by chance, which suggests that many satellites were captured into a resonance because their orbits changed after they formed. This could have happened when satellites migrated through the circumplanetary disk. Resonances could also have arisen after the circumplanetary disk disappeared, when tidal interactions with the planet forced the moons' orbits to expand.

Today, the Laplace resonance is responsible for heating Io, Europa, and Ganymede as these moons continuously adjust to the stresses imposed on them by gravitational tides induced by Jupiter. The heating is greatest on Io, which lies closest to Jupiter, and Io has become the most volcanically active object in the solar system as a result. Tidal heating helps to maintain an ocean of liquid water beneath Europa's thin icy crust and a deeper layer of liquid water within Ganymede. On Europa, the heating has broken the icy crust into a pattern of plates that are continually shifting with respect to one another. Geological activity on these moons means that their surfaces are all much younger than that of Callisto.

THE ORIGIN OF IRREGULAR SATELLITES

The unusual orbits of the irregular satellites suggest that they had a different origin. These moons almost certainly began their lives traveling around the Sun and were captured when they passed close to one of the giant planets. As we saw in Chapter 10 when we looked at the origin of

Earth's Moon, capturing a satellite requires a way to slow it down and remove some of its energy. The irregular satellites may have been slowed by gas drag when they passed through the tenuous outer atmosphere of a giant planet, or they could have been brought to a halt by colliding with an existing moon. It is also possible that the irregular satellites were captured during a rapid rearrangement in the orbits of the planets that we will discuss in Chapter 14.

Capture wasn't the end of the story for these moons, however. Many irregular satellites occur in groups moving on similar orbits. These appear to be fragments produced by catastrophic collisions between larger moons. Recent calculations suggest that the irregular satellites have suffered more collisions and fragmentation than any other group of objects in the solar system. Today, we may be looking at a few lucky remnants from a much larger population of irregular moons that existed early in the solar system.

The satellite systems of Jupiter, Saturn, and Uranus have much in common, but Neptune's system clearly had a different kind of history. Neptune's largest satellite Triton moves on a tilted orbit traveling backward around Neptune. This means that Triton was almost certainly captured, but it is still unclear how. One plausible idea is that Triton used to be part of a binary planet, rather like the Pluto-Charon system. A close approach to Neptune in the distant past separated the pair, leaving Triton in orbit around the planet, while its partner escaped, taking some of Triton's kinetic energy with it. Today, Neptune has few regular satellites. Presumably many others once existed but they were displaced when the more massive Triton entered the system. Triton probably had a highly elongated orbit shortly after its capture, crossing the paths of most of Neptune's original moons. Triton's substantial bulk carved a trail of destruction through the Neptunian system, colliding with many smaller moons or hurling them onto unstable orbits, before Triton settled onto its current orbit.

RINGS

Galileo first observed Saturn's rings in 1610, although he never appreciated their true nature in his lifetime—Saturn and its rings appeared more like a triple planet through his telescope. It was another half century before Christiaan Huygens, using a better instrument, realized that Saturn was actually surrounded by a system of rings. Four centuries later, we know that all four giant planets have ring systems, although none of the others approaches Saturn's in magnificence. Astronomers discovered that Uranus has rings in 1977 while they were carefully observing the planet's passage in front of a star. As the star drew close to Uranus, it disappeared and reappeared multiple times before it finally vanished behind the planet. When the star emerged on the opposite side of Uranus, the same pattern was repeated in reverse, showing that the starlight was being blocked by a symmetrical system of rings rather than several moons. Two year later, the Voyager 1 spacecraft discovered rings around Jupiter, and in 1989 Voyager 2 confirmed that Neptune also has rings.

Planetary rings are not solid objects but billions of separate particles the size of pebbles or boulders, each moving on its own orbit around the planet. Particles within a ring tend to be densely packed, so collisions are common. Collision speeds are typically low, so these impacts don't cause much damage to the particles. However, collisions do reduce the up and down motions of particles out of the ring plane, making the rings extremely thin as a result. Saturn's rings are thousands of kilometers (thousands of miles) across but only a few meters (several feet) thick.

Rings are usually located close to a planet, within the Roche radius—the distance at which a body with no internal strength would be torn apart by the planet's gravitational pull. Large satellites held together by their own gravity cannot survive this close to the planet. We do see some small moonlets embedded within the rings, and the material within these objects must be strong enough to prevent them being pulled apart.

All four ring systems orbit their parent planet in the same direction that the planet rotates. This suggests that the rings could be debris left over from the formation of the planets and their moons, or that the rings formed later when one or more regular satellites broke apart. If the ring material had been captured from elsewhere, it is unlikely that all four systems would orbit in the same direction.

Rings evolve over time. Collisions between ring particles gradually remove energy and cause the rings to spread apart, with some material moving toward the planet while other particles move away from it. Gravitational interactions with nearby moonlets can shepherd the ring particles and slow the spreading but not prevent it altogether. The ring particles are also continually bombarded by meteoroids orbiting around the Sun that happen to pass close to the planet. These collisions erode the rings and gradually reduce their mass over time. Rings can also be replenished when impacts onto satellites inject small pieces of debris into orbit. This replenishment may have been enough to maintain the ring systems of Jupiter, Uranus, and Neptune in their current form for the age of the solar system, but probably not the massive rings of Saturn.

Saturn's rings are truly exceptional (Figure 12.5). They are more extensive and much more massive than those of any other planet. The ring particles also have an unusual composition—almost pure water ice,

Figure 12.5. Detailed structure in Saturn's B ring imaged by the Cassini spacecraft in 2009. The image includes spokes (the ghostly radial markings near the center) and the shadow of Saturn's moon Mimas, which is a dark streak near the bottom. (NASA/JPL/Space Science Institute)

which is why they appear so much brighter than the other ring systems. Over time we would expect Saturn's rings to lose mass and become contaminated with darker, rocky material from passing meteoroids. This might mean that Saturn's rings are relatively youthful, perhaps forming as recently as 100 million years ago, much less than the age of the solar system.

Scientists have proposed several possible origins for Saturn's rings, including the breakup of a large moon in a catastrophic collision. However, such a collision is unlikely to produce rings made of nearly pure ice because all the moons we know of in the solar system contain sizeable amounts of rock. Recently, planetary scientist Robin Canup proposed a better idea. Perhaps Saturn once had two large moons like Titan—the one we see today—and a second large moon located closer to Saturn. If the inner moon formed rapidly, it would have grown hot and melted, forming a rocky core covered by a mantle of nearly pure ice. After it formed, the inner Moon could have migrated inward, passing inside Saturn's Roche limit. Saturn's gravitational tides would have stripped away the icy mantle, leaving a rocky core that continued migrating inward until it fell into the planet. The icy mantle was left behind, broken into myriad tiny pieces by Saturn's gravity, forming a set of massive rings around the planet, as well as some of Saturn's ice-rich moons. The rings have been slowly evolving and losing material ever since, leaving the more modest but still substantial system we see today.

This scenario seems plausible, but it remains unproven, and some scientists doubt that debris from the breakup of a single large moon could have gone on to form both the rings and Saturn's icy moons. For the time being, the origin of Saturn's rings remains a mystery. Perhaps their formation required a highly unusual set of circumstances, which explains why Saturn's rings far outshine those of its planetary neighbors. We may be uniquely privileged to live in a solar system graced by such a rare and beautiful phenomenon.

THIRTEEN

WHAT HAPPENED TO THE ASTEROID BELT?

THE ASTEROID BELT TODAY

Picture the asteroid belt in your mind. What does it look like? A popular image is a dense cloud of churning debris, angular boulders, and mountain-sized chunks of rock flying to and fro as far as the eye can see. Objects continually collide with one another, knocking off shards or shattering into a shower of fragments. It seems almost impossible that a spacecraft could traverse the asteroid belt and survive intact.

In some ways, this picture is correct, but it differs from reality in one striking respect: the asteroid belt between Mars and Jupiter is almost entirely empty space. Astronomers have discovered hundreds of thousands of asteroids, yet the immense size of the asteroid belt means that each object typically lies thousands of kilometers (thousands of miles) from its nearest neighbor. More than half a dozen spacecraft have passed through the asteroid belt unscathed. Asteroids are so thinly scattered that a spacecraft could travel within the asteroid belt for a million years without bumping into anything more substantial than a modest pebble.

Although little is known about most of the individual members of the asteroid belt, we have a good idea how much material there is altogether. We can gauge the masses of the biggest asteroids by watching how their gravitational tugs alter the paths of other objects that pass nearby. The sizes of smaller asteroids can be estimated by measuring their brightness. Ceres, the largest asteroid (also classed as a dwarf planet), is roughly 952 km (592 miles) in diameter. Vesta and Pallas are a little over half that size. Several hundred other asteroids are larger than

100 km (60 miles). Even so, adding up all of the known asteroids yields a remarkably modest amount of material: a total of only 0.0005 Earth masses. If all the asteroids were merged into a single body, it would be several times *smaller* than the Moon. Many small asteroids have surely escaped detection. Nevertheless, the total mass of these objects cannot amount to much since the combined gravitational pull of the known asteroids is already enough to explain the motion of nearby planets such as Mars.

There are good reasons to think that the asteroid belt was much more massive in the distant past. For a start, the section of the solar nebula between Mars and Jupiter probably contained several Earth masses of rocky material. This is at least a thousand times more than we see in the asteroid belt now. Furthermore, we know from studying meteorites that the asteroids grew to their current size in only a few million years. This rapid growth would have been impossible unless a lot more solid material was packed into the asteroid belt than there is today.

Astronomers have come up with several possible scenarios to explain how the asteroid belt lost its initial bulk. These ideas fall into two camps. One possibility is that billions of years of collisions have ground down most of the asteroids into dust, and this dust has either fallen into the Sun or has been blown out of the solar system by solar radiation. The other school of thought imagines that many asteroids were pulled out of the asteroid belt and flung into the Sun or interstellar space by gravitational perturbations from other bodies in the solar system.

GROUND DOWN BY COLLISIONS?

Asteroids certainly collide with one another from time to time; otherwise there would be no meteorites. However, collisions appear to be too infrequent to remove several Earths worth of solid material from the asteroid belt. One way we know this is by studying Vesta, the parent body of the HED meteorites. These meteorites tell us that Vesta was once hot enough to melt and differentiate, forming layers of different density. As Vesta cooled and solidified, a thin crust of lightweight basaltic rocks formed on its surface, similar to the rocks in Earth's crust.

Table 13.1. Properties of selected asteroids

Asteroid	*Location*	*Average distance from Sun (AU)*	*Average diameter or dimensions (km)*	*Density (water = 1)*	*Shape*	*Classification*
Ceres	Main belt	2.77	952	2.1	Roughly spherical	C or G
Pallas	Main belt	2.77	544	2.8	Flattened sphere	B
Vesta	Main belt	2.36	530	3.5	Flattened sphere	V
Lutetia	Main belt	2.45	$121 \times 101 \times 75$	3.4	Irregular	C or M
Mathilde	Main belt	2.65	$66 \times 48 \times 46$	1.3	Irregular, several giant craters	C
Ida	Main belt	2.86	$54 \times 24 \times 15$	2.6	Irregular, elongated	S
Gaspra	Main belt	2.21	$18 \times 11 \times 9$	2.7	Irregular, elongated	S
Eros	Near Earth/ Mars crossing	1.46	$34 \times 11 \times 11$	2.7	Irregular, elongated	S
Itokawa	Near Earth/ Mars crossing	1.32	$0.5 \times 0.3 \times 0.2$	1.9	Irregular, elongated	S

Today, Vesta's crust remains largely intact apart from one very large crater, called Rheasilvia, near the asteroid's south pole (Figure 13.1). The impact that formed this crater punched through the crust to a depth of 13 km (8 miles), exposing the rocks beneath. It seems highly unlikely that Vesta's crust would have survived with just a single major scar if most other asteroids were pulverized into dust. In fact, computer simulations show that Vesta's crust could have survived only if the asteroid belt's mass has remained small for most of the history of the solar system.

Another line of evidence comes from the radiometric clocks used to study the history of meteorites. Some of these clocks are reset whenever an asteroid suffers a violent collision because heat generated in the collision melts rocks, mixing them together and allowing gases trapped inside the rocks to escape. Many meteorites contain a record of collisions that happened when the planets were forming 4.5 billion years ago. More collisions occurred during the late heavy bombardment about 3.9 billion years ago, when impacts also formed large basins and numerous craters on the Moon, Mars, and Mercury. Apart from these brief periods, the asteroid belt seems to have enjoyed a relatively quiet history—certainly not the continual cascade of catastrophic collisions needed to turn most asteroids into dust.

It seems that collisions by themselves could not have reduced the asteroid belt to the sparsely populated one we see today. Instead, scientists have turned to the idea that many asteroids were removed by

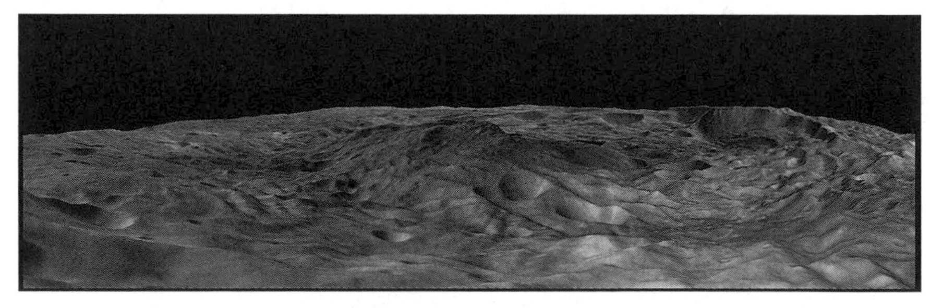

Figure 13.1. The impact basin Rheasilvia in Vesta's southern hemisphere. This perspective view from the crater rim was generated from data obtained by the Dawn spacecraft in 2012. The central mountain is about 180 km (110 miles) wide and 20–25 km (12–15 miles) high. (NASA/JPL-Caltech/UCLA/MPS/DLR/IDA/PSI)

gravitational perturbations from the planets or other objects in the solar system. Several scenarios have been proposed. Each seeks to explain both the loss of mass from the asteroid belt and the orbital characteristics of the surviving asteroids.

EMPTIED BY GRAVITY?

The orbits of most asteroids are elongated and somewhat tilted compared to those of the planets. This suggests that something pulled on the asteroids after they formed. The arrangement of different types of asteroid within the belt also points to an ancient disturbance of some kind. As we saw in Chapter 5, asteroids can be divided into different classes based on their spectra. Astronomers think each group formed at a different distance from the Sun, ending up with different properties and compositions as a result. The dry, stony S-types probably formed in the hottest region, closest to the Sun. Dark, water-bearing C-types formed in the midsection of the belt where temperatures were cooler and some water ice could form. More primitive P- and D-type asteroids formed in the outer asteroid belt. If asteroids had remained where they formed, we would expect the different asteroid types to be arranged in concentric bands like rings on an archery target. Instead, the different groups intermingle to a significant degree, suggesting that many asteroids were scattered across the belt after they formed.

Could the same perturbations that mixed the asteroids together, and changed the shape of their orbits, have forced other asteroids out of the belt altogether? In Chapter 5, we saw one way this could happen. If an asteroid is in a resonance with Jupiter, its orbit will become more elongated over time. Ultimately, the asteroid will fall into the Sun or approach Jupiter, at which point the giant planet's gravity will fling it out of the solar system. All this takes place in a million years or so—the blink of an eye compared to the age of the solar system. Any object unfortunate enough to be born in a resonance would have been lost long ago. However, resonances occupy only a few narrow zones within the asteroid belt, so resonances alone cannot explain why so many asteroids have disappeared.

Resonances would have had a much more profound effect if something caused them to move. The location of a resonance depends on the orbits and masses of other bodies in the solar system. Today, the only large objects orbiting the Sun are the major planets, and their orbits and masses are essentially constant. This means that the resonances are also fixed in place now. However, the young solar system contained another massive object—the solar nebula from which the planets formed. The combined gravitational pull of all the gas and dust in the nebula would have been enough to move some of the resonances to a different location. As the solar nebula dissipated, the resonances would have moved too, sweeping across the asteroid belt until they ended up where they are today. In the process, the sweeping resonances could have forced many asteroids out of the belt, like a rake clearing leaves from a lawn.

On the face of it, this seems like a promising idea. Unfortunately, the devil is in the detail. When scientists examined sweeping resonances using computer simulations, they found that they could explain the loss of many objects from the asteroid belt, or the tilted and elongated orbits of the surviving asteroids, or the jumbling of different asteroid types. But sweeping resonances cannot explain all these things at once. It seems that the gravitational pull of the Sun, the planets, and the solar nebula was not enough to produce the modern asteroid belt. Could there be another missing ingredient?

In the early 1990s, planetary scientist George Wetherill began to look at the problem from a different perspective. What if nothing special had happened in the asteroid belt, at least initially? Suppose planets had started to grow between Mars and Jupiter just as they did elsewhere in the solar system. To begin with, the growth of microscopic dust grains into planetesimals and larger objects would have followed the same pattern as elsewhere. Eventually, planetary embryos would have formed—objects thousands of kilometers (thousands of miles) in diameter, substantially *bigger* than Ceres, the largest asteroid. At some point, events must have intervened to prevent the appearance of fully formed planets. Suspicion falls again on the giant planets.

When Jupiter and Saturn formed, resonances associated with these planets appeared in the asteroid belt. A small fraction of the planetary embryos and planetesimals in the belt immediately found themselves on unstable orbits and were lost. This was not the end of the story, however. Gravitational tugs from planetary embryos would have been strong enough to alter the orbits of their neighbors. As a result, embryos and planetesimals moved to and fro in a haphazard manner. Sooner or later, most of these objects entered a resonance and disappeared from the asteroid belt. Computer simulations show the most likely outcome to be the loss of all the planetary embryos together with almost all the planetesimals. A tiny fraction of the planetesimals remained, moving on inclined, elongated orbits, often far away from their place of origin. These lucky survivors became the asteroids that form the belt we have today.

This scenario seems plausible, but perhaps something even more dramatic happened. In Chapter 9, we encountered the Grand Tack hypothesis in which Jupiter traversed the asteroid belt twice early in the solar system. This idea was originally proposed to explain the low mass of Mars, since Jupiter's gravity would have removed many planetesimals from the region where Mars is growing. However, the presence of Jupiter would have caused an even greater disruption in the asteroid belt as it traveled through it twice. Computer simulations of the Grand Tack scenario suggest the overwhelming majority of asteroids would have been flung out of the asteroid belt, while the survivors were jumbled together and pulled onto tilted, elongated orbits.

It is still unclear which of these ideas is correct, but it seems that the giant planets and their gravity were largely responsible for disrupting the growth of planets in the asteroid belt, and for scattering the survivors onto their current orbits. In contrast, Earth and Venus were relatively unaffected by the giant planets and their resonances, and they were free to grow into large planets moving sedately on nearly circular orbits.

ASTEROID FAMILIES

Collisions may not have been the most important factor in sculpting the asteroid belt, but they have left their mark all the same. This first became clear a century ago. The orbits of asteroids change slowly over time, undergoing small oscillations that alter the orbit's tilt and elongation. In 1918, the Japanese astronomer Kiyotsugu Hirayama realized he could get a much better picture of how the asteroids are arranged if he averaged out these oscillations. When Hirayama did this, he noticed three clusters of asteroids with very similar orbits each containing more than a dozen members. Each of these asteroid families, he decided, must have formed when a large object broke apart in a collision, producing a cloud of fragments moving in almost the same orbit as the progenitor.

Hirayama's conjecture proved to be correct. Since his discovery, astronomers have found hundreds more members of Hirayama's three families, and they have identified dozens of other families as well. It's likely that up to one-third of the known asteroids, small ones in particular, are members of families.

Asteroid families slowly disperse and disappear over time for a variety of reasons. Gravitational perturbations from the planets gradually change the orbits of some family members, while collisions destroy others. However, the most important factor is the Yarkovsky effect, which we encountered in Chapter 5. Over millions of years, the absorption and emission of heat from the Sun changes an asteroid's orbit, causing it to drift slowly across the asteroid belt. The strength of the Yarkovsky effect depends on an asteroid's size and how fast it rotates, so different asteroids drift at different rates. When an asteroid family first forms, the members are tightly clustered together in one region of the asteroid belt. Over hundreds of millions of years, the family gradually disperses in different directions until its members become indistinguishable within the asteroid belt as a whole.

Scientists can use computers to turn back the clock and work out when some families formed by calculating the motion of the family members. When doing this, there comes a time when the members all converge at a single point, the site of the collision that formed the family. For example, the asteroids in the Karin family, named after its largest member, can be traced back to a collision that happened about 6 million years ago. In this case, the parent asteroid was itself a member of the much larger Koronis family that formed several billion years ago.

Unfortunately, working backward in time becomes increasingly difficult for older families because it is impossible to say precisely how fast small asteroids drift apart due to the Yarkovsky effect. For example, the orbits of the Gefion family suggest this family formed roughly half a billion years ago when a 150-km (100-mile)-diameter asteroid broke apart, but it is hard to be more precise than this.

Luckily, some additional detective work has helped to pin down exactly when this particular family formed. Workers at a limestone quarry in Sweden have found a group of fossil meteorites, called L chondrites, which landed on Earth 467 million years ago. Other L chondrites from the same parent asteroid had their radioactive clocks reset by a large impact about 470 million years ago. This means the quarry meteorites took only a few million years to reach Earth. The only way they could have arrived here so quickly is if they formed in a collision that happened right next to a resonance. The only large family that lies next to a resonance and matches the appearance of the L chondrites is the Gefion family, so we can be pretty sure that this family formed in a spectacular collision 470 million years ago.

Another large asteroid broke up around 160 million years ago, forming the Baptistina family and generating a more prolonged shower of impacts and meteorites on Earth. One of these impacts may have caused the extinction of the dinosaurs and many other species 65 million years ago. Even the biggest asteroids can have families. Vesta, the second most massive asteroid, is accompanied by a family of fragments formed in an ancient collision, perhaps the same collision that formed the large crater Rheasilvia near Vesta's south pole.

THE MISSING MANTLE PROBLEM

In Chapter 5 we saw that meteorites display a wide range of physical and chemical properties, reflecting those of their parent asteroids. These asteroids are much too diverse to be fragments of a single planet that broke apart. However, many meteorites do belong to groups that share almost identical compositions and presumably come from a single parent body. More than 20 groups come from asteroids that never melted or only partially melted, meaning that these asteroids formed relatively late when most radioactive materials had decayed away. There are also several groups of iron meteorites from different asteroids that grew hot enough to melt. Dozens more meteorites do not belong to any group.

These appear to be unique samples from additional parent bodies. Altogether, we have samples of perhaps a hundred different parent asteroids in our meteorite collection. But the asteroid belt contains hundreds of thousands of asteroids. Why do we have samples from only a few of these?

Computer simulations that model the evolution of the asteroid belt tell us that most objects larger than 100 km (60 miles) are probably primordial—that is, they have survived more or less intact for the age of the solar system. These asteroids may be covered in impact craters, but no impact was energetic enough to break them apart completely. On the other hand, the great majority of asteroids smaller than about 100 km (60 miles) are actually fragments from the catastrophic breakup of a few larger bodies. All these fragments have the same composition as their parent asteroid. Since there are only a few hundred asteroids larger than 100 km (60 miles), we should expect to see no more than a few hundred types of meteorite at most.

In practice, there could be gaps in our collection for two reasons. Many carbonaceous meteorites are fragile, and only a fraction of these are likely to survive passage through Earth's atmosphere. Perhaps rocks from some asteroids are more fragile still and never make it to the ground intact. Second, rocks from some parent bodies are more likely to land on Earth than others. Asteroids that lie in the inner asteroid belt, or close to a resonance, or bodies that have recently broken up into many pieces will contribute more meteorites than other asteroids. This probably explains why about 80 percent of all meteorites come originally from just three asteroids—the parent bodies of the three types of ordinary chondrite.

Scientists believe that iron meteorites come from the cores of asteroids that melted. But what happened to the corresponding rocky material that formed the mantles of these bodies? A few asteroids have spectra that match those of mantle rocks, but they are very rare. Some nonmetallic meteorites come from asteroids that have partially or wholly melted, but these do not match the minerals we would expect to see in the missing mantles of the iron parent bodies. These exotic meteorites must come from some other kind of parent body instead.

The rarity of mantle rocks in our meteorite collection and in the asteroid belt, known as the "missing mantle problem," is a long-standing puzzle. There are several reasons why iron fragments might survive better than rocky fragments when asteroids break apart. Iron lies in the core of a differentiated asteroid, while rocky material lies near the surface. Thus, rocky material will be the first to be removed when an asteroid is bombarded, while iron is the last to be exposed. As a result, rocky fragments have to survive in space for longer than iron ones. Most of the rocky mantle may be peeled away in small fragments—chips from the surface—while the iron core remains as a single piece, making it harder to disrupt later. Last and most important, iron is much stronger than rock: a piece of iron is likely to survive in the asteroid belt at least 10 times longer than a rocky fragment of the same size.

If most differentiated bodies broke apart early in the solar system, perhaps all the mantle material has been ground down to dust and lost over the billions of years since then. This would mean that intact differentiated asteroids are very rare in the asteroid belt today. Perhaps Vesta and a handful of others are all that remain.

However, collisional erosion cannot be the whole story. Primitive asteroids, the parent bodies of chondritic meteorites, are no stronger than the mantle rocks from differentiated asteroids. How did so many primitive asteroids survive when almost none of the differentiated ones did? Part of the explanation may simply be that differentiated bodies were relatively rare to begin with and none have survived. Still, if almost all differentiated bodies were destroyed in violent collisions, how did Vesta survive with only a single large crater on its surface?

Astronomer William Bottke and his colleagues recently came up with a possible explanation: perhaps the parent bodies of the iron meteorites formed closer to the Sun, in the region that now contains the terrestrial planets. Objects would have been more tightly packed nearer the Sun, so collisions would have been more frequent than in the asteroid belt. Many, perhaps most, differentiated bodies were disrupted by violent collisions. Gravitational perturbations from larger bodies scattered some of these fragments into the asteroid belt. Both iron and rocky fragments arrived in the asteroid belt, but only the stronger iron objects have survived for the age of the solar system. Later on, the parent bodies of primitive meteorites formed in the asteroid belt. Most of these objects survived, leaving an asteroid belt today that is a mixture of intact primitive bodies and fragments of iron.

ASTEROIDS REVEALED AS WORLDS

In the past three decades, our view of asteroids has improved dramatically. This transformation began in the late 1980s and 1990s when astronomers began directing radar beams at several asteroids as they passed close to Earth. By examining how the beam was reflected back to Earth from different parts of the asteroid, they obtained the first 3D representations of asteroids, showing their shapes in some detail. It was immediately obvious that these asteroids were not smooth, round, or nearly round like planets. Instead, the asteroids were often highly elongated or irregular objects. Their surfaces were covered in bulges, ridges, and depressions. Some even looked like dumbbells, as if two separate objects were nudged up against each other, held together by gravity.

The picture became clearer still when spacecraft began to take close-up pictures of asteroids. In the 1990s, the Galileo spacecraft flew close by two S-type asteroids, Gaspra and Ida, on its way to Jupiter (Figure 13.2). Both asteroids turned out to be angular, elongated objects with numerous impact craters and fractures on their surface. Galileo also discovered that Ida has a small moon, Dactyl, roughly 1.5 km (1 mile) in diameter. It was the first time a satellite had been found orbiting an asteroid.

With hindsight, it shouldn't have been too surprising that some asteroids have moons. Violent collisions between asteroids can generate a huge number of fragments, all of which attract one another due to their gravity. Recent computer simulations show that many fragments are likely to reaccumulate into larger objects and that some of them end up in orbit about one another. Since Dactyl's discovery, astronomers using telescopes have found many other asteroid satellites.

The first space mission dedicated to studying an asteroid was NASA's NEAR (Near Earth Asteroid Rendezvous) Shoemaker mission, launched in 1996. The spacecraft's main target was Eros, an S-type near-Earth

Figure 13.2. Ida and Dactyl. This image of the main-belt asteroid Ida and its moon Dactyl was taken by the Galileo spacecraft in 1993. It was the first conclusive evidence that asteroids could have moons. Ida is 56 km (35 miles) long and Dactyl only 1.4 km (4,600 ft) across. (NASA/JPL)

asteroid. Eros's orbit lies relatively close to Earth, but the limited amount of fuel available meant that NEAR had to take a roundabout route, moving out into the asteroid belt then back past Earth again, using our planet's gravity to fling the spacecraft toward Eros.

Along the way, NEAR flew past another asteroid, Mathilde, giving us our first view of a C-type asteroid. Mathilde is a strange object indeed. Although very roughly spherical, its surface is marred by five enormous impact craters, like giant dimples in a golf ball. Any one of these impacts should have been enough to shatter Mathilde and disperse the pieces if the asteroid is a solid object. The fact that Mathilde survived these collisions suggests that it is actually a "rubble pile"—a jumbled assortment of pieces held loosely together by gravity. Somewhat surprisingly, computer simulations show that a rubble pile is harder to disrupt than a solid body, since much of the energy of a collision moves the existing pieces around and heats them up rather than flinging fragments into space. Mathilde has a remarkably low density, less than half that of solid rock and only slightly greater than water. Much of the asteroid's interior must consist of empty space—gaps between the rubble that the asteroid's gravity is too feeble to squeeze shut.

NEAR finally arrived at Eros and went into orbit on Valentine's Day 2000. Eros proved to be a banana-shaped object roughly 35 km (22 miles) in length with several prominent craters (Figure 13.3). NEAR spent the next year carefully mapping the surface and measuring the asteroid's composition. Eros appears to be a solid body rather than a rubble pile, but its surface is strewn with boulders ejected in a large impact. The asteroid is also covered in a layer of dust and small rocks that may be tens of meters (many tens of feet) thick. In a few places, the dust forms "ponds," extremely flat surfaces where dust has been shaken loose by an impact and gently settled into depressions (Figure 13.4).

In a bold decision, NASA decided to end the mission by bringing the spacecraft down onto the surface of Eros as gently as possible even though it was not designed to land. This was the first ever landing on an asteroid. After several maneuvers, NEAR touched down safely, relaying a series of close-up pictures during the descent. Once on the surface, sensors measured the asteroid's composition in detail and beamed the results back to Earth for the next two weeks. The measurements showed

Figure 13.3. The near-Earth asteroid Eros. This mosaic of Eros's southern hemisphere was taken by the NEAR Shoemaker spacecraft on November 30, 2000. Eros is 33 km (21 miles) long. (NASA/JPL/JHUAPL)

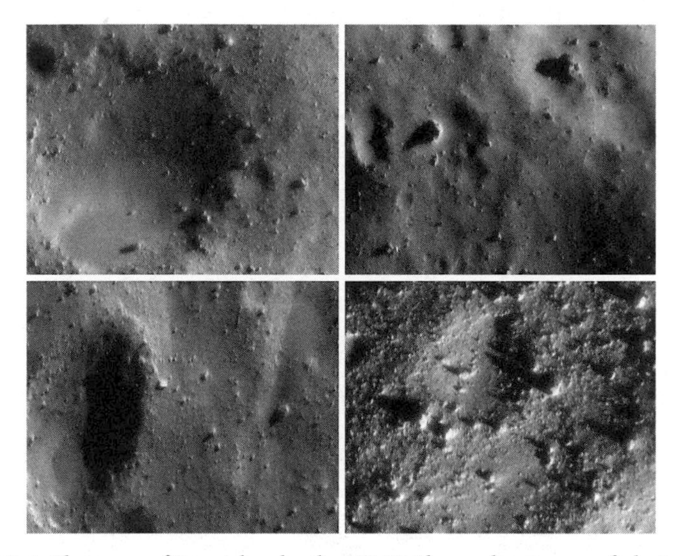

Figure 13.4. Close-ups of Eros taken by the NEAR Shoemaker spacecraft during low-altitude passes in 2001. *From upper left to lower right*: the images show Eros's boulder-strewn surface at increasing resolution. The two top scenes are about 550 m (1,815 feet) across and the bottom two about 230 m (760 feet) across. (NASA/JPL/JHUAPL)

that Eros has a composition more like ordinary chondrites than any other type of meteorites, a strong hint that these meteorites come from S-type asteroids. This discovery was confirmed 10 years later by the Hayabusa mission.

The Japanese Space Agency launched Hayabusa in 2003 with the goal of returning our first sample from an asteroid. Two years later, Hayabusa arrived at its target, the main-belt asteroid Itokawa. The spacecraft took high-resolution pictures and other data that showed the asteroid is a rubble pile, like Mathilde, with very few craters on its surface. During its rendezvous, Hayabusa landed on the surface twice, but the mechanism designed to collect a sample failed to operate properly on both occasions. Luckily, the spacecraft's maneuvering dislodged some fine grains of dust and lofted these into the vessel's sampling horn. Despite a series of malfunctions, Hayabusa successfully brought a precious cargo of 1,534 tiny particles to Earth. Analysis of the dust suggests that Itokawa is made up of fragments from the interior of a larger asteroid. The degree of space weathering indicates that the surface dust has been exposed to space for about 8 million years, so Itokawa probably formed in a relatively recent collision.

To date, spacecraft have approached eight asteroids closely enough to acquire detailed pictures. One of the most unusual of these is Lutetia, an irregular asteroid with a diameter of 100 km (60 miles) visited by the Rosetta spacecraft in 2010 (Figure 13.5). Based on its spectra, astronomers have classified Lutetia as belonging to the class of M-type asteroids, which are often thought to be the parent bodies of iron meteorites.

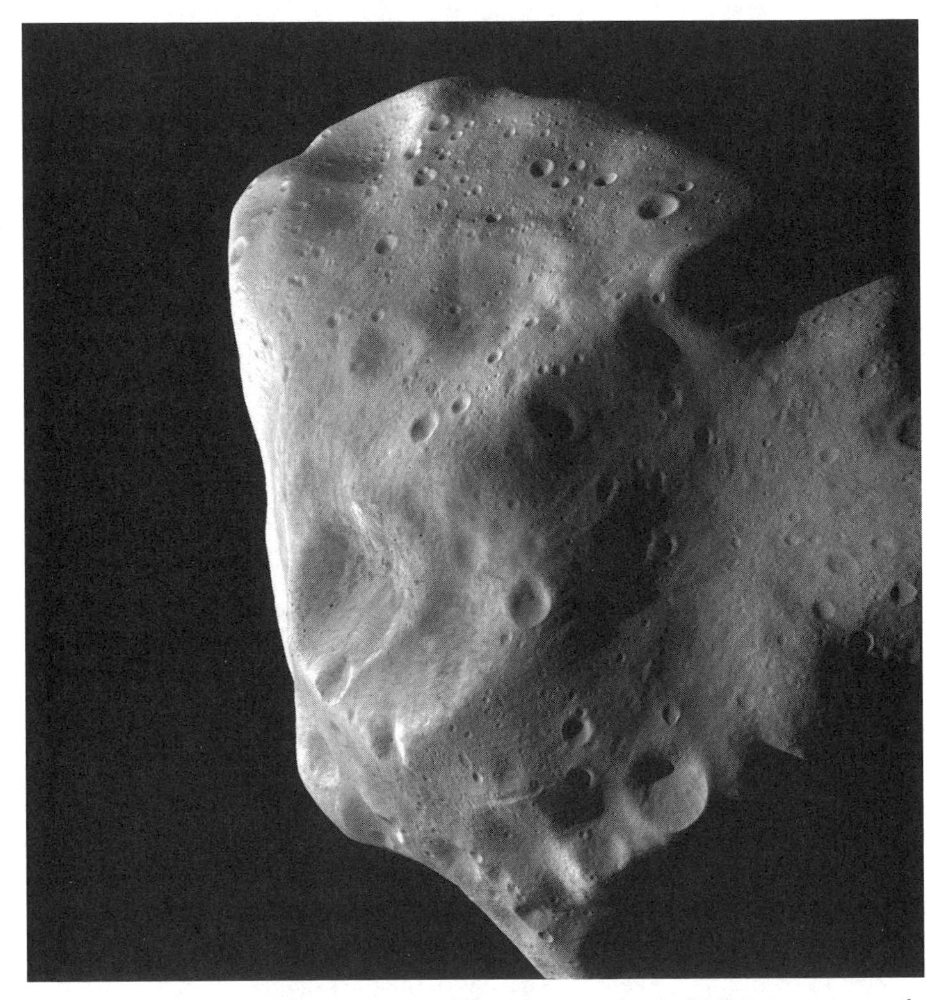

Figure 13.5. The main-belt asteroid Lutetia. This image was taken by ESA's Rosetta spacecraft in July 2010. Lutetia is about 100 km (62 miles) in diameter and, although now in the main belt, has a composition that suggests it was originally closer to the Sun. (ESA 2010 MPS for OSIRIS Team MPS/UPD/LAM/IAA/RSSD/INTA/UPM/DASP/IDA)

Lutetia shows no sign of metal on its surface, but its density is one of the highest known for any asteroid. This can be explained if it has an iron-rich interior. Data from Rosetta, combined with observations made using the Spitzer Space Telescope and telescopes on the ground, suggest that Lutetia could actually be the parent of the rare and exotic meteorites called enstatite chondrites, which we met in Chapter 5.

All the asteroids surveyed at close range so far appear to be dry, rocky bodies. This is very likely to change in 2015 when the Dawn mission arrives at the largest asteroid, Ceres. The spectrum of Ceres shows that its surface contains hydrated, clay-like minerals, while the asteroid's density and gravity suggest it has a thick shell of water ice over a denser, rocky core. Ceres could even contain more water than all the fresh water on Earth, as well as a tenuous atmosphere of water vapor.

In Chapter 1, we saw how a few small bodies in the outer asteroid belt sometimes develop comae and tails of gas and dust like a comet. These objects, like Ceres, must contain substantial amounts of water in the form of ice in their interiors, protected by an insulating blanket of dust. Recent collisions could have exposed the ice to sunlight, causing it to evaporate and blow off a cloud of dust every time the asteroid gets close enough to the Sun. These objects seem quite unlike most other asteroids. They probably have more in common with comets, the icy dust balls that occasionally visit us from the frigid outer extremes of the solar system. In the next chapter, we will look at these mysterious objects in more detail.

FOURTEEN

THE OUTERMOST SOLAR SYSTEM

WHERE DO COMETS COME FROM?

Until the mid-20th century, little was known about the true nature of asteroids and comets. Asteroids looked like single points of light through a telescope, and astronomers tended to think of them as miniature planets, albeit ones with somewhat more elliptical and inclined orbits. The great majority of known asteroids moved within the main asteroid belt between Mars and Jupiter, but astronomers were aware that a few came closer to Earth or traveled beyond Jupiter. Comets seemed to constitute a separate family, distinct from asteroids. They grew much brighter as they approached the Sun, becoming enormously extended objects with a diffuse coma and one or more tails that could extend for millions of kilometers (millions of miles). Typically, comets followed highly elongated paths that could be inclined at any angle to the orbits of the planets. One group, the Jupiter-family comets, moved mostly or entirely within the planetary system, but others, called long-period comets, traveled far beyond the realm of the planets.

The most puzzling aspect of comets was how they had managed to survive for the age of the solar system. Comets had such low masses that they couldn't be measured, yet they shed large amounts of material each time they passed close to the Sun to form their coma and tails. Occasionally, comets disintegrated completely, leaving behind only a stream of dusty debris. Surely, all the solar system's original complement of comets would have lost their material or disintegrated long ago if they had been traveling on their current orbits the whole time.

Comets could be recent interlopers captured from interstellar space rather than part of the original fabric of the solar system, but this seemed unlikely since comets are so numerous. In 1931, the Estonian astronomer Ernst Öpik came up with a different explanation. Perhaps comets have been a part of the solar system since the beginning, but they spend the great majority of their lives lurking far from the Sun, lying dormant and too faint to see. Occasionally, comets could escape from this distant reservoir, arriving in the inner solar system to replace other comets that have disappeared over time. This idea seemed promising, but Öpik had no way to test it.

In 1950, the Dutch astronomer Jan Oort examined the matter again. Oort noted that many long-period comets travel on extremely elongated orbits that take them tens of thousands of AU from the Sun. He realized that these objects would remain bound to the solar system by the Sun's gravity, if only very loosely. This meant the comets were part of the solar system rather than outsiders. Most of the long-period comets observed by astronomers seemed to be passing near the Sun for the first time. Oort deduced that there must be a vast spherical swarm of dormant comets orbiting far from the Sun, which he thought was roughly 50,000 to 150,000 AU away. Every once in a while, the gravity of a passing star or the Milky Way as a whole, pulled one of these icy bodies out of the distant swarm and sent it hurtling toward the Sun, where it began to form a coma and tails and became visible as a comet.

Oort's original study was based on the limited amount of data available to him—relatively few long-period comets were known at the time—but his idea has been confirmed as more comets have been discovered and computer simulations have calculated the trajectories of comets over millions of years. The distant swarm of comets is now known as the Oort cloud. It remains far beyond the reach of the most powerful telescopes in use today, but astronomers accept that it must exist, forming the outermost part of the solar system.

Around the same time that Oort was writing about the orbits of comets, American astronomer Fred Whipple came up with a convincing explanation for their physical nature. Astronomers had measured the spectra of comets and deduced that their tails and comae contain atoms of carbon, hydrogen, oxygen, and nitrogen paired together in various combinations. These seemed to be broken fragments of common molecules such as water, methane, ammonia, and carbon dioxide that form ices when they are very cold. Whipple later wrote:

> It became obvious to me in the late 1940s that comets must carry a large reservoir of these parent molecules to keep comets active for hundreds, or even possibly thousands, of revolutions about the Sun. In addition, some comets must be big enough and solid enough to graze the Sun without experiencing total destruction. The answer was clear: The nucleus of a comet must be a great mass of ices embedded with dust or meteoric particles—in other words, it must be a huge, dirty snowball.

When a solid cometary nucleus is heated by the Sun, the ices begin to evaporate, producing streams of gas that escape from the surface carrying rocky dust grains with them (Figure 14.1). It was an elegant explanation for why many comets develop two distinct tails—one made of gas and one of dust. Whipple published his idea in 1950, and it quickly caught on.

The next question was where these dirty snowballs formed in the first place. Clearly, comets must have formed in a very cold environment, which pointed to the outermost reaches of the solar system. However, it seemed unlikely that the solar nebula that gave rise to the planets could have extended as far out as the Oort cloud. Even if the solar nebula had been that large, its matter would have been spread so thinly in the Oort cloud that enough of it could never have come together in one place to form a comet. Comets must have begun their lives somewhere other than the Oort cloud.

In 1951, the Dutch-American planetary scientist Gerard Kuiper reasoned that a population of small, icy bodies should have formed in the solar nebula just beyond Neptune. These objects, he suggested, were later dispersed far and wide by gravitational perturbations from the planets, forming the Oort cloud as a result. If Kuiper were correct, the region beyond Neptune would be empty today except for Pluto, which, at the time, was thought to be a large planet. Other researchers saw things somewhat differently. One was the Irish engineer, economist, and astronomer Kenneth Edgeworth, who wrote in general terms about the

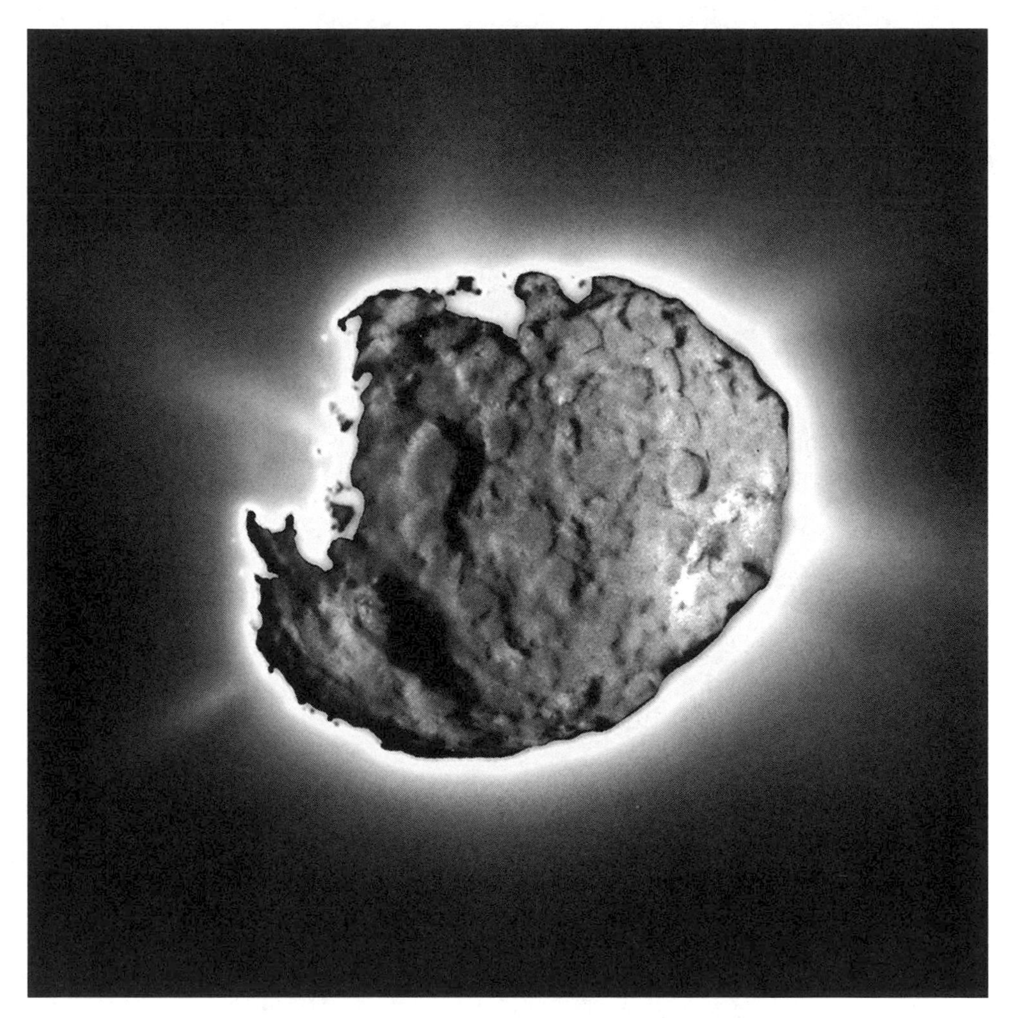

Figure 14.1. A composite image of Comet Wild 2, consisting of two exposures taken 10 seconds apart by the Stardust spacecraft in January 2004. A short exposure showing surface details on the comet's nucleus is superimposed on a long exposure showing jets of gas and dust streaming from the very active surface. (NASA/JPL-Caltech)

origin of the solar system in the 1940s. In his view, any plausible formation scenario would produce a solar system that gradually petered out with distance from the Sun, leaving a region beyond Neptune that "is in fact a vast reservoir of potential comets."

CENTAURS

It would be decades before astronomers possessed instruments powerful enough to see whether there really were any small objects beyond Neptune. However, a tantalizing hint of things to come came in 1977. In that year, Charles Kowal was conducting a survey of the solar system, looking for unusual objects in much the same way that Clyde Tombaugh had done 50 years earlier when hunting for Pluto. In October, Kowal discovered an object moving very slowly across the sky. Later he wrote:

> From its motion, I was able to deduce that this object must be near the orbit of Uranus. Yet Uranus and its satellites were on the other side of the sky. I realized that nothing else big enough to be seen was known at this distance! After more photographs were taken it became possible to compute an orbit. This confirmed that the object was nearly 18 AU from the Sun at the time of discovery. Its perihelion distance [closest approach to the Sun] was only 8.5 AU. Clearly the object was no ordinary asteroid.

Here was an object moving entirely in the outer solar system that spent most of its time between Saturn and Uranus. It never came anywhere near to the asteroid belt, and its orbit was more like that of a comet than a planet. Given its brightness, the newly discovered body had to be around 200 km (120 miles) across, 10 times bigger than any comet nucleus but 10 times smaller than any planet. As Kowal said, it simply did not fit any pigeonhole. Kowal named his object Chiron, after the most famous of the mythological centaurs, the son of Saturn and the grandson of Uranus. Chiron was also designated asteroid number 2060. "Now, for the people who like to label things," he said, "We can say that Chiron is a Centaurian asteroid!"

This was not the end of the story. In 1988, as Chiron moved nearer to the Sun, its brightness increased by 75 percent and it developed a hazy coma and a faint tail. Chiron had become a comet. This object was not going to remain in the pigeonhole created for it. Scrambling to keep up, astronomers reclassified Chiron, making it both a comet and an asteroid at the same time. The distinction between comets and asteroids was becoming decidedly fuzzy, and Kowal warned that the existing terminology was inexact. "Some 'asteroids' may, in fact, be inactive comets," he said, and "one object called an asteroid may be totally different from another object called an asteroid." A second object, Pholus, was found moving on a similar orbit in January 1992, followed by dozens more "Centaurs," as they came to be known.

LOOKING BEYOND NEPTUNE

By 1980, computers were powerful enough for scientists to calculate how the orbits of Oort cloud comets changed over thousands or millions of years. These simulations showed that it was very unlikely that most Jupiter-family comets came from the Oort cloud. Unlike long-period comets, the orbits of Jupiter-family comets are aligned quite closely with the orbits of the planets. The simulations showed that it was impossible to start with a spherical distribution of comets like the Oort cloud and turn it into a much flatter arrangement like the Jupiter-family comets.

To address this problem, Uruguayan astronomer Julio Fernández speculated that there could be a flattened belt of comets between about 35 and 50 AU from the Sun. He showed that such a belt could supply Jupiter-family comets fast enough to account for the number we see in the inner solar system today. Fernández emphasized that his comet belt was not a substitute for the Oort cloud, but supplementary to it. If he was correct, the solar system actually contained two large reservoirs of dormant comets rather than one.

Fernández was not optimistic that his belt of comets would be discovered any time soon. These hypothetical objects would be so faint that they could barely be detected with the best instruments available at the time. Fortunately that was about to change. The 1980s saw the advent of sensitive CCD detectors that allowed astronomers to peer farther into the outer solar system than ever before.

On September 14, 1992, the International Astronomical Union issued electronic circular number 5611, one of a series of frequent messages that provide details of routine astronomical discoveries. This particular circular was anything but routine:

> *1992 QB1* D. Jewitt, University of Hawaii, and J. Luu, University of California at Berkeley, report the discovery of a very faint object with very slow (3″/hour) retrograde near-opposition motion, detected in CCD images obtained with the University of Hawaii's 2.2-m telescope at Mauna Kea. . . . Some solutions are compatible with membership in the supposed "Kuiper Belt," . . . but a satisfactory definition of the orbit will clearly require follow-up through the end of the year.

The wording was cautious since the orbit of a distant body can be difficult to pin down at first. However, further observations proved beyond doubt that the new object, called 1992 QB1, lay beyond Neptune, precisely where Edgeworth, Kuiper, and Fernández had predicted a belt of primitive bodies should exist. The discoverers, David Jewitt and Jane Luu, said that their search had been "motivated by the desire to understand the apparent emptiness of the outer solar system," echoing the conviction of Franz Xaver von Zach and other searchers two centuries ago that the space between Mars and Jupiter could not be empty. Jewitt and Luu had begun their long and difficult systematic search six years earlier. With the discovery of 1992 QB1, their persistence had finally paid off.

THE KUIPER BELT

More discoveries soon followed. By 2010, astronomers had found over a thousand similar bodies orbiting beyond Neptune, which we now call trans-Neptunian objects (TNOs). The most distant lies 100 AU from the Sun, 3 times farther out than Pluto. As more objects were discovered, patterns began to emerge in the size, shape, and orientation of their orbits. Each group must have experienced a different history that can tell us something about the formation and early history of the solar system.

Table 14.1. Properties of selected trans-Neptunian objects

Object	*Location*	*Average distance to Sun (AU)*	*Average diameter (km)*	*Comments*
Pluto	Kuiper belt	39	2,300	5 known moons, thin atmosphere
Makemake	Kuiper belt	45	1,500	
Haumea	Kuiper belt	43	1,400	Highly elongated shape
1992 QB1	Kuiper belt	44	160	First discovered since Pluto
Eris	Scattered disk	68	2,300	One known moon
Sedna	Scattered disk	544	1,000	Reddish in color

Many TNOs, including 1992 QB1, belong to what is now known as the "classical Kuiper belt." These objects are also familiarly called "cubewanos," a name contrived from their prototype's designation (Q-B-one-o). Classical Kuiper-belt objects lie sufficiently far from Neptune that they are only weakly affected by the planet's powerful gravity. This makes their orbits reasonably stable for billions of years. Neptune has a nearly circular orbit 30 AU from the Sun, while classical Kuiper-belt objects mostly occupy a band between 42 and 48 AU. The orbit of 1992 QB1 is fairly typical: slightly elliptical, ranging between 40.9 AU and 46.6 AU from the Sun, and inclined by 2 degrees compared to Earth's orbit.

Some members of the Kuiper belt lie outside the classical belt, but they manage to avoid coming close to Neptune because they lie in a stable resonance. A prime example is Pluto. Pluto takes about 249 years to orbit the Sun while Neptune takes 165 years, so that their orbital periods are in the ratio 2:3. Even though Pluto's orbit crosses that of Neptune, the two bodies are never remotely in danger of colliding. Every time Pluto crosses Neptune's orbit, the larger planet lies on the far side of the Sun, so a collision is impossible. Pluto actually gets nearer to Uranus than it ever does to Neptune.

Over 100 known objects share the 2:3 resonance with Pluto, and collectively these objects are called "Plutinos." Another 100 or so objects are protected by other resonances with Neptune such as 3:5, 4:7, and

1:2. Unlike the unstable resonances that form the Kirkwood gaps in the asteroid belt, which we encountered in Chapter 5, the resonances in the Kuiper belt actually help to keep their occupants away from trouble. Resonances, it turns out, are subtle phenomena, each one behaving in a different way, which makes it hard to predict their effect without studying each in detail. The stable resonances beyond Neptune may have played an important role in the Kuiper belt's history, as we will see shortly.

The classical Kuiper belt and resonant populations occupy a doughnut-shaped region of space between about 30 and 50 AU from the Sun. Beyond 50 AU, the number of objects drops sharply. Careful studies reveal that this boundary is real and not simply caused by the difficulty of spotting more distant objects. The classical Kuiper belt ends rather abruptly at 50 AU rather than slowly petering out, as Edgeworth imagined. Curiously, this outer boundary coincides with the 2:1 resonance with Neptune, and this may not be an accident.

The region beyond 50 AU is not completely empty. Astronomers have found more than 100 objects farther from the Sun, but based on their orbits, these clearly belong to a different component of the Kuiper belt. The largest object found so far is Eris, about 2,400 km (1,500 miles) in diameter, similar in size to Pluto and 27 percent more massive. Eris's orbit is very different from that of 1992 QB1 and other members of the classical Kuiper belt. The orbit is highly elongated and tilted at an angle of 44 degrees to Earth's orbit. As Eris goes around its 557-year orbit, its distance from the Sun ranges between 38 and 98 AU. Eris is currently almost at the farthest point of its range, making Eris and its moon Dysnomia the most distant visible objects in the solar system.

Eris, and other bodies with similar orbits, belong to what astronomers call the "scattered disk." Unlike the classical Kuiper belt, objects in the scattered disk occasionally come close enough to Neptune to experience a strong gravitational tug. These tugs have altered their orbits, making them highly elliptical and inclined. On their closest approach to the Sun, scattered disk objects mingle with other members of the Kuiper belt, but their distinctive orbits give them away. The Centaurs, which orbit mainly between Saturn and Neptune, were presumably scattered in the same way, except that Neptune's gravity flung them inward toward the Sun rather than outward into the scattered disk. It is now generally believed that the scattered disk is the source of most Jupiter-family comets, and that Centaurs are currently making the same transition from TNOs to visible comets.

SEDNA

Neptune's gravitational tugs have sculpted the scattered disk, giving its members their characteristically tilted and elongated orbits, but there are some objects beyond even Neptune's reach. Just over a dozen TNOs are known to follow elongated orbits between about 35 and 100 AU from the Sun. These objects never come close enough to Neptune for the planet's gravity to affect them significantly. Astronomers call these "detached objects," and how they arrived in their present orbits remains something of a puzzle. The most promising explanation for most of them is that they came under Neptune's influence in the past but have subsequently had their orbits changed by the effects of resonances.

One detached object has an orbit so bizarre that it stands apart from the others. This is Sedna, which we first met in Chapter 7. Sedna travels on a highly elliptical orbit that brings it no closer to the Sun than 76 AU, and carries it out to 960 AU. No part of its path passes through the Kuiper belt. When Michael Brown, Chad Trujillo, and David Rabinowitz discovered Sedna in 2003, they wrote:

> The orbit of this object is unlike any other known in the solar system. . . . The only mechanism for placing the object into this orbit requires either perturbation by planets yet to be seen in the solar system or forces beyond the solar system.

Sedna could not have been flung onto its distant orbit by Neptune, and the mechanism that might explain the nearer detached objects does not work for Sedna. At the same time, Sedna lies too close to the Sun to be part of the Oort cloud as we usually think of it. Brown and his colleagues suggested that Sedna could belong to a smaller "inner Oort cloud," offering three possible explanations for how it arrived in its present orbit.

One idea is that Sedna was pulled onto its current path by an undiscovered Earth-sized planet orbiting about 70 AU from the Sun. This seems highly unlikely, however—astronomers would almost certainly have found a planet this large by now, and we would see its gravitational influence on the orbits of other TNOs if it existed. A second scenario imagines that a star passed close to the solar system sometime within the past few billion years, perturbing Sedna onto its unusual orbit. Such events are very rare however, so the odds are very much against this. The most plausible suggestion is an idea we introduced in Chapter 7: that Sedna was set on its current orbit when the Sun was still part of the close-knit cluster of stars in which it formed. Stars in a cluster repeatedly encounter one another, typically passing much closer than stars outside a cluster. In such an environment, objects on the periphery of the burgeoning solar system would have experienced stronger and more frequent gravitational tugs than they do today, making it much more likely that some objects would end up on Sedna-like orbits.

It would be much easier to work out which scenario is correct if we could study a large population of objects similar to Sedna rather than a single body. When Brown and his colleagues discovered Sedna, they imagined it would only be a matter of time before its cousins turned up. "Study of these populations will lead to a new knowledge of the earliest history of the formation of the solar system," they wrote. To date, astronomers have found just two other TNOs that are remotely comparable to Sedna. Neither body has an orbit as far removed from Neptune, and these objects do little to help distinguish between theories for Sedna's orbit and origin. Alas, Sedna remains stubbornly unique, just as Pluto did for over 60 years.

THE NATURE OF TRANS-NEPTUNIAN OBJECTS

We know relatively little about what TNOs are made of, and even less about how their compositions came about. Most TNOs are too faint to have their spectra analyzed in any detail apart from assessing their overall color. The few TNOs that are bright enough to yield a useful spectrum show a clear dichotomy: all the large bodies display signatures of ices, while the smaller ones do not—nobody knows why.

TNOs come in a surprisingly diverse variety of hues. Unfortunately, it is hard to say whether these different colors mean that the objects formed out of different materials or whether they were similar to begin with and evolved in different ways. The color of an object can change over time when radiation transforms simple ices such as water, methane, nitrogen, and ammonia into complex organic compounds that tend to have a reddish hue. Over billions of years, an icy body can develop a dark, red-tinged crust covering a more pristine interior. Occasionally, fresh ices may spew out from the interior making parts of the surface bright and white again, complicating the picture. Collisions can also form fresh surfaces and expose deep materials in objects that have differentiated into layers.

Of all the TNOs, Pluto and its largest moon Charon have been studied in the most detail. The densities of Pluto and Charon suggest they have a similar composition overall—roughly one-third water ice and two-thirds rocky minerals, with smaller amounts of other materials. The two bodies look very different, however. Pluto's surface is a patchwork of light, dark, and reddish-orange areas (Figure 14.2). Much of the surface is coated with frozen nitrogen and methane, which have partially vaporized to create a thin atmosphere. Charon's surface is more uniform than Pluto's, is gray rather than red, and seems to be dominated by water ice. These two worlds are separated by a mere 20,000 km (12,000 miles), yet they are conspicuously different—perhaps we will find out why when the New Horizons spacecraft visits Pluto and Charon in 2015.

A striking feature of the TNO population is that many objects have moons. Dozens of moons have been discovered, and nearly 10 percent of TNOs that have been studied closely have at least one companion. In addition to Charon, Pluto is known to have four smaller satellites. Nix and Hydra, with diameters of roughly 50 km (30 miles), were discovered with the Hubble Space Telescope in 2005. Two smaller moons were found in 2011 and 2012, and have yet to be named.

If a TNO has a moon, it is possible to calculate the pair's combined mass by observing the moon's orbit and applying one of Kepler's laws.

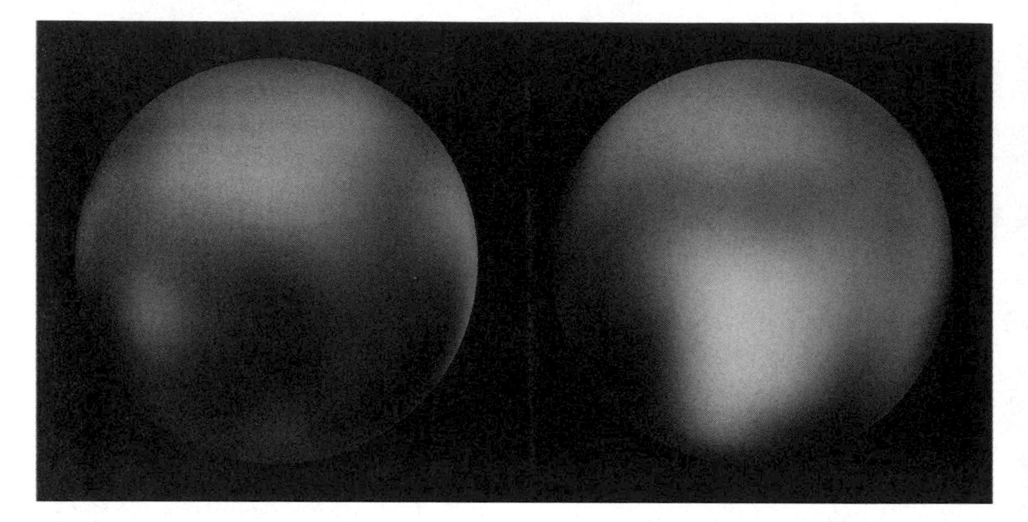

Figure 14.2. Maps of Pluto's surface. These maps of two opposing hemispheres of Pluto were released in 2010. They were created by processing images obtained by the Hubble Space Telescope in 2002–3 and are likely to remain the best available images of Pluto until the arrival of the New Horizons mission in 2015. The white areas are frost and the dark areas a carbon-rich residue deposited when ultraviolet radiation from the Sun breaks up the methane on Pluto's surface. A comparison between these maps and images taken in 1994 shows that Pluto's surface had changed significantly since 1994. (NASA, ESA and M. Buie [Southwest Research Institute])

This is how we know the masses of Pluto and Eris so well. In general, it is much harder to measure a TNO's size. Doing so relies on the detection of faint infrared emissions or, failing that, an educated guess at how strongly the object reflects sunlight, which might be wrong by as much as a factor of 10. Putting together the mass and diameter of an object gives us the density. The limited data available suggest that TNOs have a wide range of densities, between about 0.5 and 3 g/cm^3, compared to values of about 1 and 3 g/cm^3 for pure water ice and rock, respectively. Pluto lies in the middle at 2.03 g/cm^3, while the average for the small number of comet nuclei that have had their densities estimated is around 0.6 g/cm^3. Some of these objects, those with densities below 1 g/cm^3, are likely to be porous with large amounts of empty space in their interior.

After observing the region beyond Neptune for two decades, it is clear that the population of TNOs is complex and varied. The trans-Neptunian population contains many large, icy bodies that are fairly dense and resemble miniature planets. Jupiter-family comets, which were once members of the Kuiper belt, tend to be small, fragile, low-density objects by comparison, consisting of loosely bound grains of dust and ice.

At one time, most scientists imagined comets to be pristine, icy fossils preserving the most primitive material to be found anywhere in the solar system. It turns out that comets are more complicated, and cometary material is more highly processed, than anybody expected. Two recent space missions helped to bring about this transformation. In 2005, the Deep Impact spacecraft launched a large copper cylinder that crashed at high speed into comet Tempel 1. Immediately after the collision, the comet released an immense cloud of gas and dust from its interior. The dust included crystalline silicates that could have been produced only by intense heating, confirming a puzzling discovery made earlier by studying the spectra of comets. The dust from Tempel 1 also contained clays and carbonates, which suggest liquid water existed inside the comet at some point in the past.

Samples of dust collected from comet Wild 2 by the Stardust spacecraft, and returned to Earth in 2006, also contained crystalline silicates (Figure 14.3). Overall, the dust from Wild 2 and Tempel 1 contains

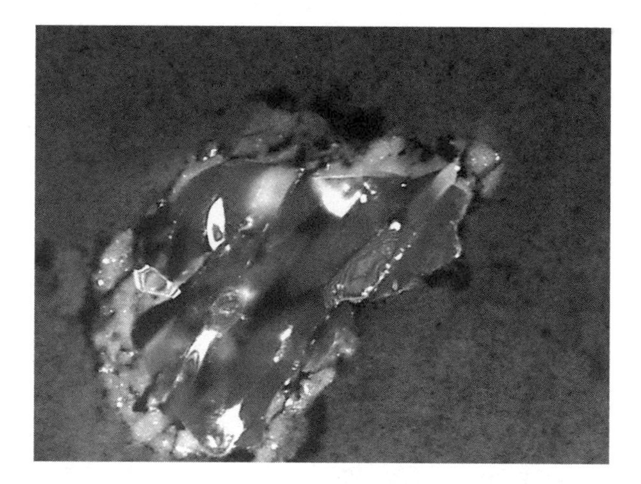

Figure 14.3. A dust particle collected by the Stardust spacecraft during its mission to Comet Wild 2. The particle is about 2 micrometers across and is made up of the silicate mineral forsterite, also known as peridot in its gem form. It is surrounded by a thin rim of melted aerogel, the substance used to collect the comet dust samples. (NASA/JPL-Caltech/University of Washington)

chemical elements in the same proportions as they are present in the Sun, so these comets are essentially made of the same stuff as the rest of the solar system. However, by the time these comets formed in the solar nebula, many dust grains had been thermally and chemically processed and thoroughly mixed up. The comets must have been assembled in the frigid outer extremes of the solar system in order to preserve their icy constituents, but they also swept up material that had been processed much nearer to the Sun in furnace-like temperatures. Clearly, the building blocks of comets moved around a good deal within the solar nebula, and this may have been true for the material that built the other bodies in the solar system as well.

WHERE HAVE ALL THE PLUTOS GONE?

The population of icy planetesimals that once existed in the outer solar system has changed almost beyond recognition in becoming the modern Kuiper belt. One reason to reach this conclusion is that the total mass of the Kuiper belt today is very low, amounting to less than 10 percent of Earth's mass. If material had been this sparse in the outer regions of the solar nebula, it is difficult to see how objects as large as Pluto could have ever formed. In fact, plausible models for the solar nebula suggest that the Kuiper belt once contained some 300 times more solid material than it does today.

Neptune's moon Triton, and Pluto's moon Charon, provide further clues to a more populous Kuiper belt in the past. Triton is similar in size, density, and surface composition to Pluto (Figure 14.4). As we saw in Chapter 12, Triton travels backward around Neptune on an orbit that is tilted with respect to its planet's equator—the only large satellite in the solar system that moves on a retrograde orbit like this. Triton's unusual orbit strongly suggests that it was captured by Neptune. Prior to this, Triton would have orbited the Sun, like Pluto. Charon, by contrast probably formed when Pluto was hit by another large body—an event somewhat similar to the giant impact that gave rise to Earth's Moon.

On the face of it, Triton and Charon formed in very different ways, but their origins had one thing in common: each involved a very close

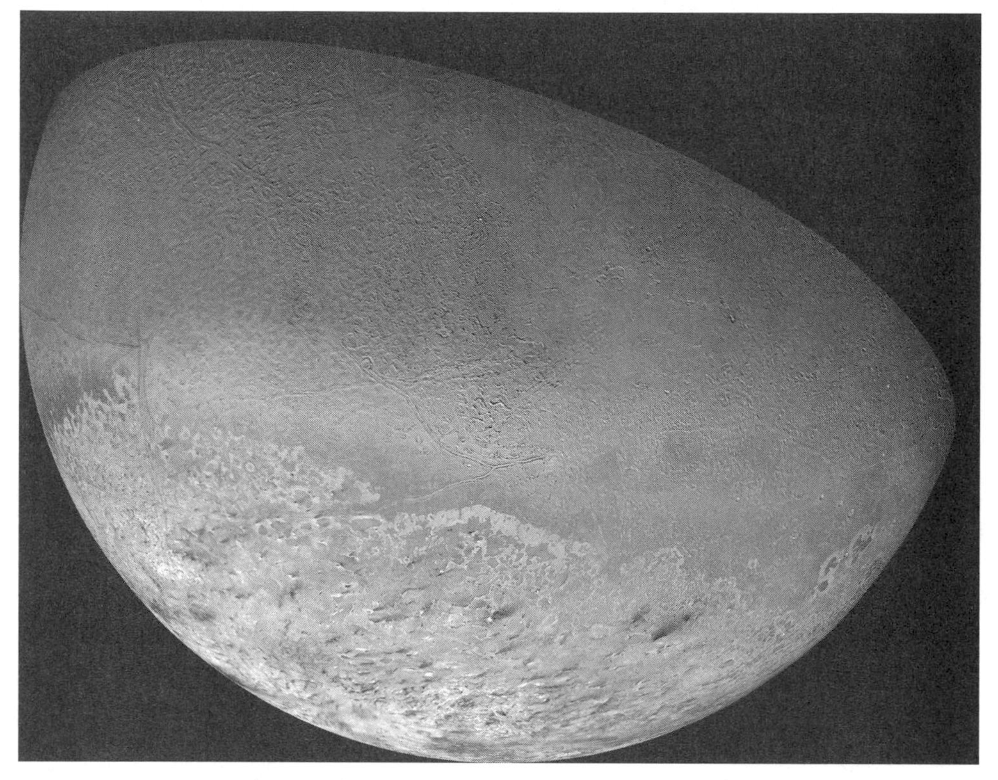

Figure 14.4. A mosaic of Neptune's moon Triton taken in 1989 by Voyager 2 during its flyby of the Neptune system. It is likely that Triton is a captured Kuiper belt object. If so, the Voyager images of Triton are the best close-up images of a Kuiper belt object until the New Horizons mission arrives at the Pluto system in 2015. Triton has the coldest surface known anywhere in the solar system (about –235°C or –391°F). It is so cold that most of Triton's nitrogen is condensed as frost, making it the only satellite in the solar system known to have a surface made mainly of nitrogen ice. The vast south polar cap is believed to contain methane ice. The dark streaks overlying this ice are believed to be icy and perhaps carbonaceous dust deposited from huge geyser-like plumes, some of which were active during the Voyager 2 flyby. (NASA/JPL/USGS)

encounter between two objects moving through the vast expanse of the outer solar system. Such encounters are incredibly unlikely today because the region is so thinly populated. To Alan Stern, the lead scientist on NASA's New Horizons mission, Triton and Charon are "smoking guns"—compelling evidence that many more icy worlds once existed beyond Neptune. If he is right, it raises the question, "Where have all the Plutos gone?"

As we saw in Chapter 8, the solar nebula probably contained millions of small, solid objects when the planets were forming. The sizes of these planetesimals were roughly in the range 1 to 1,000 km (1 to 500 miles) or more. Some planetesimals initially lay beyond Neptune, but many others traveled between the orbits of the growing planets. The planets swept up a substantial fraction of these planetesimals, but many of those in the outer solar nebula were thrown aside following a close encounter with one of the giant planets. Many of those thrown outward would have ended up in the Oort cloud as a result.

Initially, the orbits of the giant planets were probably a good deal closer together than they are today, as we saw in Chapter 12. These orbits changed over time as the giant planets traded energy with passing planetesimals. Computer simulations show that Jupiter must have moved toward the Sun over time as it gave up some of its energy to hurl planetesimals out to the Oort cloud. Somewhat surprisingly, Saturn, Uranus, and Neptune all moved outward—in the opposite direction to Jupiter. The weaker gravity of these planets meant they were actually better at nudging planetesimals inward toward Jupiter than they were at flinging them toward the Oort cloud, so these three planets gained energy and moved away from the Sun as a result.

Soon after the first members of the Kuiper belt were discovered, planetary scientist Renu Malhotra suggested that we might still see a signature of planets' migration today. As Neptune moved away from the Sun, some objects found themselves captured into resonances, including Pluto and the Plutinos. These objects moved outward in lockstep with Neptune traveling ahead of the planet. Neptune's gravitational tugs would have forced the orbits of the Plutinos to become more elliptical the longer they remained in a resonance, and this would explain why many of the Plutinos, including Pluto itself, have quite elongated orbits today.

Other planetesimals never entered the safety of a resonance, or did so only briefly, and Neptune's gravity would have quickly removed them, depleting the trans-Neptunian region of much of its initial mass. Some of these objects ended up in the scattered disk where the survivors remain today. Others now lie in the Oort cloud or were thrown out of the solar system altogether.

Eventually, Neptune stopped migrating. Presumably this happened when Neptune reached the outer edge of the solar nebula, which must have been at about 30 AU—Neptune's distance from the Sun today. Originally, the region beyond 30 AU was empty, and it is now populated by objects scattered there while Neptune was migrating outward. If this scenario is correct, the classical Kuiper belt we see today is entirely composed of objects that formed somewhat closer to the Sun. Beyond the 2:1 resonance, Neptune's influence tailed off dramatically, and this resonance marks the outer boundary of the classical Kuiper belt as a result.

THE NICE MODEL

This scenario sounds reasonable, but is it the whole story? Could the slow outward migration of Neptune really be responsible for the wholesale dismemberment of the primordial Kuiper belt and the loss of more than 99 percent of the Plutos that once existed there? In 2005, a group of researchers proposed a much bolder scenario in which the orbits of the outer planets underwent a dramatic rearrangement long after the solar system had finished forming. This idea was the brainchild of an international team of specialists in planetary dynamics: Rodney Gomes, Harold Levison, Alessandro Morbidelli, and Kelomenis Tsiganis. It has come to be known as the "Nice model" because the research was based at the observatory in the French city of Nice.

The aim of the Nice model was to find a unified scenario for the evolution of the solar system after the planets had formed, accounting for the origin of the various components of the Kuiper belt, the Trojan asteroids, the irregular satellites of the giant planets, and even the occurrence of the late heavy bombardment in the inner solar system that we described in Chapter 10. The Nice team tested their ideas using an extensive suite of computer simulations and compared the results with the solar system we observe now.

According to the Nice model, the four giant planets formed on nearly circular orbits between about 5.5 and 17 AU from the Sun—all interior to Uranus's present orbit. The most promising computer simulations actually place Neptune's original orbit closer to the Sun than Uranus's, but otherwise the order of the planets was the same as today. A substantial belt of icy planetesimals, totaling about 35 Earth masses, occupied the region between 15 and 35 AU from the Sun—that is, from the orbit of the outermost planet to slightly beyond Neptune's current orbit. Interactions between planetesimals on the inner fringes of this belt caused Uranus, Neptune, and Saturn to move outward, while the planetesimals were scattered inward. When Jupiter encountered planetesimals, its stronger gravity was able to catapult the small objects outward to the Oort cloud or beyond. As a result, Jupiter moved slightly toward the Sun. To this extent at least, the Nice model resembles the conventional picture we described earlier.

According to the Nice model, however, the scattering of icy planetesimals and slow adjustment of the planetary orbits continued for some 600 million years until Jupiter and Saturn found themselves in an unstable resonance with each other. The consequences of this resonance for both of the planets, and for the solar system as a whole, were sudden and highly disruptive. In a relatively short time, the orbits of Jupiter and Saturn became much more elliptical. Thanks to its elongated orbit, Saturn came close to Uranus and Neptune, and its gravitational tugs transformed their nearly circular orbits into extended, elliptical ones as well. Neptune became the most distant planet at this stage, after swapping positions with Uranus.

On their new, elongated orbits, the three outermost giant planets penetrated the main part of the icy planetesimal belt. Vast numbers of planetesimals were scattered out of the belt, many in the direction of the inner solar system. This rain of planetesimals was responsible for many of the impact craters we see on the ancient surfaces of the Moon, Mars, and Mercury. The changing orbits of the giant planets also opened up new stable niches in the solar system, and these were soon occupied by the flood of displaced planetesimals, giving rise to the Trojan asteroids and irregular satellites of the giant planets. Eventually, only about 1 percent of the planetesimals remained in the belt beyond the planets. After this dramatic episode, the orbits of Jupiter and Saturn continued to evolve until they were no longer in a resonance. Further gravitational interactions with the surviving planetesimals allowed the orbits of the giant planets to become nearly circular again, and deposited the planets at their current distances from the Sun. The migration process naturally petered out when Neptune neared the outer edge of the planetesimal belt and nearly all its members were ejected.

We can never be completely sure that a model like this tells us what actually happened in the past, but the Nice model has a number of aspects that make it plausible and appealing. It explains why the giant planets were able to form within the lifetime of the solar nebula—because they formed closer to the Sun—and how they came to occupy their current orbits. The Nice model simulations do a good job of reproducing the main properties of the Kuiper belt, scattered disk, irregular satellites, and Trojan asteroids. The model can also explain why there was a long, relatively quiet, gap between the formation of the planets, and the flurry of impacts on the Moon and inner planets that formed the late heavy bombardment.

The wholesale scattering of icy planetesimals associated with the Nice model may explain the characteristics of some of the objects that occupy the outer parts of the main asteroid belt and the Trojans that share an orbit with Jupiter. Many of these are D-type asteroids, characterized by a dark reddish color and relatively featureless spectrum, which seem to have more in common with comet nuclei than rocky asteroids. Some main-belt asteroids also develop a coma and a tail like a comet from time to time, as we saw in Chapter 1. It seems plausible that these objects formed in the outer solar system and were later implanted in the asteroid belt during a planetary upheaval like that envisioned by the Nice model.

By successfully reproducing many features of the modern solar system, the Nice model has won considerable support, but that does not necessarily mean it is correct. Like many previous theories, it may be fatally undermined by future discoveries. For example, planetary scientists recently realized that the nearly circular orbits of the terrestrial planets place strong constraints on how the giant planets could have changed in the past—the orbits of the giant planets probably evolved in a series of discrete jumps rather than smoothly over time, otherwise the inner planets would have been forced onto more elliptical orbits than we see today.

For the time being, however, the unified way in which the Nice model explains several apparently unrelated phenomena makes it attractive. Even if the Nice model turns out to be incorrect, it seems very likely that the distribution of small bodies in the Kuiper belt and elsewhere in the outer solar system was modified to a large degree by the gravitational perturbations and migration of the giant planets early in the history of the solar system.

FIFTEEN

EPILOGUE: PARADIGMS, PROBLEMS, AND PREDICTIONS

Worlds on worlds are rolling ever
From creation to decay
Like the bubbles on a river
Sparkling, bursting, borne away.

—*Percy Bysshe Shelley, Hellas*

In 2010, NASA announced its latest science plan. One of the key goals for NASA's future planetary science program is to learn how the Sun's family began and how it has changed over time. This is a direct response to the kind of questions that scientists and the public keep asking: where do we come from, and how did the world come to be the way it is? Scientists and engineers around the globe are pursuing this goal with every tool at their disposal. Astronomers and space agencies in dozens of countries are helping us to see the solar system as never before, transforming points of light into real worlds, and even bringing samples of those worlds back to Earth. At the same time, the stunning discovery of hundreds of other planetary systems in our galaxy has provided a powerful stimulus to understand how planetary systems form and evolve, and to find out what makes one system different from another.

The rapid pace of recent developments makes now a good time to take stock of what we know, even though the story is still incomplete.

In this book we have journeyed back in time to see how early scientists and philosophers tried to make sense of our planet and its neighbors, and how their ideas have evolved over time. We also traveled back 4.5 billion years to the earliest days of the solar system, piecing together the story of its formation based on the painstaking efforts of thousands of astronomers, geologists, physicists, and chemists. In this final chapter, we summarize the picture as we see it today and look ahead to what we might learn in the coming years. Finally, we peer far into the future to speculate about the ultimate fate of the solar system.

THE PARADIGM: SOLAR SYSTEM EVOLUTION IN A NUTSHELL

Two lines of evidence—radiometric dating of rocks and studies of the solar interior using helioseismology—tell us that the Sun and the solar system formed at about the same time, some 4.5 billion years ago. Earth, Mars, the Moon, and the asteroids all formed within the space of 100 million years, and it seems likely that the other planets formed at this time as well.

The solar system began as a rotating, disk-shaped cloud of gas and fine dust grains. The shape of this cloud explains why the planets all orbit the Sun in the same direction today and why their orbits almost lie in the same plane. Astronomers see similar protoplanetary disks around many young stars today. These disks are roughly similar in size to the solar system and contain about the right amount of material to build a system of planets. Disks can't last long because they are seen only around stars less than a few million years old, but this is apparently enough time to start building a planetary system. Searches for planets orbiting other stars suggest that at least 20 percent of stars have planets, and many more stars have debris disks formed by asteroids or comets. It seems that the appearance of a planetary system is a natural part of the process that forms stars like the Sun.

The planets and asteroids in the solar system formed from the bottom up, starting small and growing larger over time. The starting materials were fine grains of dust and ice that we see in protoplanetary disks, as well as millimeter-sized (0.04-inch) particles, such as chondrules, which make up the bulk of many meteorites. Radiometric dating tells us that small objects like asteroids formed before the planets were fully grown and that small planets like Mars formed earlier than larger planets like Earth. The main exceptions to this rule are the gas-rich giant planets. Jupiter and its cousins must have acquired their gas from the Sun's protoplanetary disk (the solar nebula) in the few million years before the solar nebula disappeared.

Collisions played an important role in planet formation. Dust grains and small particles stuck together to form larger objects. Later, bodies became large enough for their gravity to assist further growth. Today we see evidence for collisions throughout the solar system. Giant collisions are the most plausible explanation for the origin of Earth's large Moon, the high density of Mercury, the formation of Pluto's largest satellite Charon, and possibly the striking differences between Mars's northern and southern hemispheres. On a smaller scale, the size distribution of the asteroids, irregular satellites, and Kuiper-belt objects all point to a prolonged history of catastrophic breakups due to collisions.

Collisions between planet-sized bodies and the decay of radioactive isotopes generated tremendous amounts of heat. The kind of rocks seen on Earth, the Moon, Mars, and some asteroids strongly suggest that these bodies were once hot enough to melt, either partially or completely. This melting allowed denser materials such as iron, nickel, and gold to sink to the center, forming metal-rich cores, while lighter rocky materials floated upward.

Much of the heat left over from formation remains trapped inside planets today, along with energy released by radioactive elements. This heat is gradually escaping through volcanoes on the rocky planets. Early volcanism on these planets released large amounts of carbon dioxide, water vapor, and other gases, which accumulated to form an atmosphere. On Earth, the water condensed to form oceans while most carbon dioxide reacted with rocks and was removed from the atmosphere. Venus, where temperatures were higher, lost its water to space and retains a thick carbon dioxide atmosphere. The weaker gravity of Mercury and Mars means that these planets lost most of their atmospheres at an early stage.

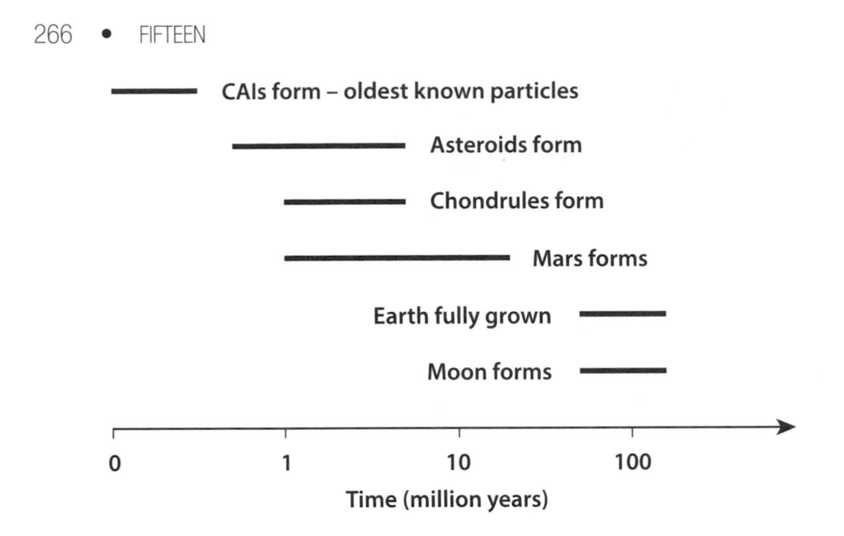

Figure 15.1. A timeline of events in the early evolution of the solar system.

Planets grew larger in the outer solar system mainly because the Sun's gravity is weaker there, which allowed planets to sweep up material from a much larger region. It seems likely that the giant planets all began life as solid bodies that became large enough for their gravity to pull in gas from the solar nebula. These planets are all enriched in rock and ice-forming elements compared to the Sun, and measurements of their gravity by spacecraft suggest that each giant planet has a dense core at its center. Gas accretion went much further on Jupiter and Saturn than on Uranus and Neptune, presumably because the solar nebula dispersed before the outer two planets could grow very large.

The solar system contains two extensive belts of minor bodies: the asteroid belt between Mars and Jupiter, and the Kuiper belt beyond Neptune. In each of these regions, either material failed to coalesce into a single planet or planets formed there and were later lost. The most likely explanation in both cases is that gravitational perturbations from the nearby giant planets frustrated planet formation. The low mass of material in the two belts and the orbital distribution of the surviving objects suggest that most objects were removed from these regions and ultimately disappeared. Some objects collided with the Sun or a planet, while others were ejected into interstellar space.

UNSOLVED PUZZLES

Today, most scientists are reasonably happy with the scenario outlined above. But there are several things we still don't understand about the origin of the solar system. It is the nature of science that our ideas will continue to change over time as new information becomes available and discoveries occasionally refuse to fit the existing paradigm. While it is always difficult to predict what lies ahead, there are some outstanding issues that may be resolved in the near future.

One of the most difficult questions to answer is how dust grains and small particles in the solar nebula grew into mountain-sized planetesimals—bodies so large that gravity could hold them together and pull in more material. This issue is particularly important because all the subsequent stages of planet formation depend upon it. Laboratory experiments, astronomical observations, and computer simulations are all being used to address the problem. Finding an answer may require a better understanding of how small particles interact with gas in a low-gravity environment. As we saw in Chapter 8, two ways that growth could happen in real protoplanetary disks have been discovered recently, and with luck, a breakthrough may be just around the corner.

At the opposite end of the scale, another vexing question is how fully formed planets interact with their protoplanetary disk. One of the great surprises when extrasolar planets were discovered was that many planets lie extremely close to their star. Theoretical models predict that these planets must have formed at greater distances and migrated inward. There seem to be two mechanisms that could make planets move close to their star like this: gravitational interactions between a planet and its protoplanetary disk and gravitational interactions between different planets. At present, we don't know which mechanism is more important, or the extent to which migration sculpted planetary systems in general and the solar system in particular.

We do know that a third kind of migration took place in the solar system, caused by gravitational interactions between the giant planets and leftover planetesimals. This migration has left its mark in the orbital arrangement of small bodies throughout the outer solar system today.

Unfortunately, the timing of this rearrangement is unclear. If it took place relatively late, it could have been the trigger for the intense episode of impacts on the Moon and inner planets that took place 3.9 billion years ago. This question is particularly intriguing because the earliest signs of life on Earth occurred right after this bombardment ended, and it seems unlikely that this is a coincidence.

The current spacing of the planetary orbits makes sense in terms of stability. If the planets were significantly closer together, they would have become unstable by now and two or more planets would have collided. What is less clear is why the planets have their current *sizes*. What made Earth and Venus so much larger than their neighbors Mars and Mercury? Why did Jupiter end up being three times more massive than Saturn, when both clearly became large enough to accrete gas from the solar nebula? We do not know the answers to these questions yet, but there are tantalizing hints that the mass of a planet depends as much on events that occurred elsewhere in the solar system as what was happening in the planet's immediate vicinity. For example, recent computer simulations suggest that if Jupiter formed before Saturn, Saturn's growth could have been permanently stunted, preventing it from ever growing as large as Jupiter itself.

SEARCHING THE SOLAR SYSTEM FOR ANSWERS

Space missions will play an important role in helping to answer these questions. Scientists have high hopes that future missions will lead to important discoveries as they have done repeatedly in the past. A good way to learn about conditions in the early solar system is to explore bodies that have changed little over the past 4.5 billion years. Asteroids are prime candidates for such missions since they finished forming even before the major planets were fully assembled. Icy bodies from the outer solar system are also obvious targets, including Kuiper belt objects and their cousins the comets, which make themselves more accessible by visiting Earth's neighborhood. None of these bodies is necessarily a pristine sample of the solar nebula. However, they are likely to preserve some materials that existed when the solar system was forming, and these bodies may bear telltale scars of events that happened long ago.

We first met the Rosetta mission in Chapter 1. Rosetta, the most ambitious mission to a comet ever attempted, was launched in 2004 and should arrive alongside Comet Churyumov-Gerasimenko in 2014 after a circuitous journey. In November 2014, Rosetta will release a lander called Philae, named after an island in the River Nile where archaeologists found an obelisk that helped to decode the Rosetta stone. Philae will anchor itself on the surface of the tiny comet, 3 by 5 km (2 by 3 miles), and transmit data to Rosetta to be relayed back to Earth. After depositing Philae, Rosetta will remain in low orbit around the comet as it travels toward its closest approach to the Sun at a distance of 1.3 AU and then begins its return trip back to the vicinity of Jupiter's orbit. If all goes to plan, Rosetta will continue to return data until December 2015.

Rosetta carries 11 scientific instruments, and Philae a further 9, designed to prod and probe the comet in every conceivable way. These instruments will measure its shape, structure, appearance, and composition, including the isotopes, chemical compounds, and minerals from which it is made. By staying with the comet for a whole year, Rosetta will watch how it reacts to the increasingly intense radiation and solar wind as the comet moves closer to the Sun in its orbit. This should help establish how much the Sun has modified the comet's composition over its lifetime, and give us a better idea of its original makeup.

Comet Churyumov-Gerasimenko orbits the Sun every 6.6 years, and its surface must have changed substantially after making many approaches to the Sun. However, Philae will be able to drill beneath the surface to analyze more pristine samples in the comet's interior. The composition of the samples will provide clues to how and where the comet formed as well as its thermal history. For example, noble gases like argon that were trapped within a comet's frigid ices when it formed can easily escape into space if the ice ever grows warm. The amount of noble gases remaining in the comet's interior will tell us how hot the comet was in the past, and how pristine it remains today.

Comet Churyumov-Gerasimenko probably formed beyond the orbit of Neptune, but it must have changed a good deal since then. The New

Horizons mission to the Kuiper belt will study icy bodies in their original home. This small mission was launched in 2006 and will have a brief encounter with Pluto and its moons as it speeds past them at a distance of about 10,000 km (6,000 miles) in July 2015. Over the following five years, the mission planners hope to redirect New Horizons to other objects in the Kuiper belt if enough fuel is available.

When the seven instruments on New Horizons return pictures and data from Pluto, it will be the first time that an icy planetesimal has been studied in detail. Previously, the nearest we have come to this is Voyager 2's observations of Neptune's moon Triton, which is thought to be a Kuiper belt object captured by Neptune. New Horizons should produce a global map of Pluto with a resolution of 1.6 km (1 mile), and images of selected areas at a resolution 30 times better—far more detailed than the best pictures we have of Triton. The distribution of craters on Pluto and its moon Charon will tell us about the history of impacts on these bodies and how the Kuiper belt region was populated in the past. New Horizons will also map the chemical composition of Pluto and Charon, helping us to understand conditions during their formation. If all goes well, this great cache of new data from the hitherto unexplored Kuiper belt will dramatically improve our knowledge of this region of the solar system and the primitive occupants to be found there.

While New Horizons continued on its 5-billion-km (3-billion-mile) journey, a spacecraft called Dawn began exploring remnants from the formation of the solar system that lie closer to home. Dawn's targets are the dwarf planet Ceres and the asteroid Vesta, both located in the main asteroid belt. Scientists chose this pair because both have probably changed little since the dawn of the solar system, yet they are dramatically different specimens. Like ancient fossils of different species, studying these objects should help us understand how we ended up with such a diverse asteroid belt today. Vesta melted early in its history and differentiated into layers like the terrestrial planets, apparently losing its water in the process. Ceres is also differentiated, yet its surface appears to be covered in water-bearing minerals, and it probably harbors vast quantities of ice or liquid water in its interior. Clearly, Vesta and Ceres had different histories, perhaps because they formed at different distances from the Sun. Ceres may represent a new kind of object, one that is halfway between the familiar rocky bodies of the inner solar system and the icy bodies that orbit far from the Sun.

A different kind of mission may help solve the long-standing puzzle of how the giant planets formed. Just as Dawn began to collect data at Vesta, a small spacecraft called Juno set off on a 5-year voyage to Jupiter carrying a suite of seven instruments. Juno will spend a year in 2016–17 observing Jupiter, trying to resolve questions left unanswered by Galileo, the previous Jupiter mission launched more than 20 years earlier. Two of Juno's main goals are to get a better estimate for the mass of Jupiter's core and to measure the composition of its atmosphere, giving us a clearer picture of what the planet is made of, especially how much water it contains. Scientists will need to know both these things before they can say whether the giant planets began their lives as solid bodies like Earth, or formed in another way.

Looking further into the future, scientists are hoping to secure substantial samples from two asteroids and return them to Earth to study in detail. In 2011 NASA gave the go-ahead for its first mission to collect a sample from an asteroid. OSIRIS REx, due to launch in 2016, will visit a small, carbonaceous near-Earth asteroid, known as 1999 RQ36. After studying the asteroid for several months it will scoop up 60 grams (2 ounces) of material from the surface and return it to Earth in 2023. Meanwhile, Japan plans to build on the partial success of its Hayabusa mission, which returned a tiny sample of dust grains from asteroid Itokawa in 2010. A second mission, Hayabusa 2, should launch in 2014, visiting the small, carbonaceous near-Earth asteroid 1999 JU3 and returning a larger sample to study in 2020.

OTHER PLANETARY SYSTEMS

Space missions can tell us much about the solar system and its formation by allowing us to examine planets, moons, comets, and asteroids at close quarters and in much greater detail than we can from Earth. But space missions are limited to the solar system. Astronomers will need to study many more planetary systems to place our own in context, and to answer one of the questions that intrigues us most: are Earth-like planets common, or did a unique set of circumstances come together to make our planet? We already know that Earth is special within our solar system. It lies at just the right distance from the Sun to maintain liquid water on its surface, unlike Venus, for example. Earth has enough water to form oceans but not so much that oceans completely cover the surface. Earth is large enough for plate tectonics to operate, keeping the atmosphere fresh and replenished, unlike Mars. At the same time, Earth is small enough that it never acquired a huge gaseous envelope like Jupiter.

Nearly all the most Earth-like extrasolar planets known by mid-2013 were discovered by the Kepler mission. Kepler was launched in 2009 specifically to look for planetary systems. The spacecraft consists of a telescope with a 1.4-meter (55-inch) primary mirror, designed to detect the tiny drop in a star's brightness when a planet crosses in front of it. Kepler continuously monitored over 100,000 stars in the constellations Cygnus, Lyra, and Draco, which lie in the direction of the Sun's motion around the galaxy. This means most of the stars in Kepler's field of view are roughly the same distance from the galactic center as the solar system and, like the Sun, close to the plane of the galaxy. There is a good chance that many of these stars formed under similar conditions as did the Sun, which hopefully increases our chances of finding planets like Earth.

Many things can cause a star's brightness to dip apart from a passing planet, so all of the possible planets found by Kepler have to be investigated in more detail by astronomers on the ground before they can be confirmed. By 2013, over 3,000 planetary candidates had been found. Many of these objects, if confirmed, would be comparable in size to Earth, and some would lie in their star's habitable zone, possibly giving them a climate similar to Earth (Figure 15.2).

Unfortunately, in May 2013, Kepler stopped collecting data after a hardware failure. Although much work remains to be done confirming candidate planets, Kepler has already increased the pace of discovery dramatically. Based on Kepler's findings so far, astronomers think that billions of stars in our galaxy host planetary systems. Thousands of stars within 1,000 light-years of the Sun are likely to have a planet in the star's habitable zone. There is every chance that planets closely resembling Earth will be found as Kepler's candidate objects are scrutinized.

Figure 15.2. Kepler-47 is one of the many planetary systems discovered by the Kepler mission. This diagram compares the Kepler-47 system (*top*), in which two planets orbit a double star, to the inner solar system (*bottom*). One of the two Kepler-47 stars is similar to the Sun, while the other is much smaller and 100 times dimmer. The planet Kepler-47 c orbits within the "habitable zone" (where liquid water could exist). However, it is thought to be a gas giant slightly larger than Neptune and so unlikely to be hospitable to life. (NASA/JPL-Caltech/Tim Pyle)

FUTURE EVOLUTION OF THE SOLAR SYSTEM

In this book, we have looked at the history of the solar system and how scientists have uncovered this history piece by piece. However, the story of the solar system is not over. The Sun and its companions will continue to evolve for billions of years. We end by gazing far into the future, to see what might be in store for Earth and its neighbors.

Some events are easy to predict. The huge number of objects in the asteroid belt and Kuiper belt means that these bodies will continue to collide with one another and break apart, slowly grinding these belts into dust. The gravitational pull of the planets will free other asteroids from their confinement, allowing them to roam farther afield until they run into a planet or escape from the solar system. Occasionally, two large asteroids will collide, producing a new family of objects and injecting a shower of fragments into orbit around the Sun, some of which will ultimately hit the planets. New comets will continue to arrive in the planetary region from the Oort cloud. Once every 100 million years or so, a big comet or asteroid will hit Earth, causing a mass extinction of the kind that wiped out a large fraction of species 65 million years ago, including the dinosaurs.

The orbits of satellites throughout the solar system will evolve slowly due to gravitational tidal interactions with their host planet. In some cases this will lead to dramatic changes. Mars's moon Phobos will break apart into a ring or collide with Mars a few tens of millions of years from now. Billions of years in the future, something similar will happen to Neptune's largest satellite Triton. The Moon, which is currently moving away from Earth, may change direction one day and eventually collide with our planet.

Even more dramatic changes may be in store. We have already seen how there is no guarantee that the orbits of the planets will remain stable forever. Astrophysicist Jacques Laskar has examined the motions of the planets for billions of years into the future using computer simulations. The evolution of planetary orbits on such long timescales cannot be predicted with certainty, but it is possible to say how likely some events are, rather like weather forecasts predict the chances of rain a few days in the future. Laskar finds that there is a small but very real possibility that the orbits of Mercury and Venus will begin to cross sometime in the future, leading to a collision. What this huge impact would mean for Earth is unclear, but it would certainly be devastating for the two planets involved.

In the long term, the most dramatic change in the solar system will happen to the Sun. As nuclear reactions in the Sun's core convert hydrogen into helium, the Sun will grow larger and hotter. This slow but inexorable trend will affect every other object in the solar system. For a while, changes in Earth's atmosphere may offset the increasing amount of sunlight, keeping our planet pleasantly habitable as it is today. At some point within the next few billion years, this compensation scheme will break down and Earth's surface will grow ever hotter. The oceans will boil, ultraviolet light from the Sun will break apart water molecules in the air, allowing hydrogen to escape to space, and Earth will come to resemble our hellish sister planet Venus.

Some 5 billion years from now, changes in the Sun will speed up dramatically. Our star will swell into a red giant, engulfing Mercury and Venus, and possibly Earth as well. These planets will spiral into the Sun's fiery interior and vaporize. However, our loss may be offset by beneficial changes elsewhere. As temperatures increase throughout the solar system, the large, icy satellites of Jupiter and Saturn may grow warm enough to develop thick atmospheres and oceans of liquid water.

Who knows, if balmy conditions last long enough, on a newly thawed world far from the Sun, the cycle of life could begin again.

absolute zero
The lowest possible temperature. It is the zero point of the Kelvin temperature scale and is equivalent to –273.15°C (–460°F). At absolute zero, a substance has no heat energy.

achondrite
A type of stony meteorite that doesn't contain chondrules. Achondrites come from asteroids that have partially or completely melted and differentiated.

albedo
The proportion of the light falling on a body or surface that is reflected, expressed as a fraction or a percentage.

alpha particle
The nucleus of a helium atom, consisting of two protons and two neutrons. Alpha particles are emitted by some radioactive nuclei.

angular momentum
The inertia an object has because it is rotating or revolving around a center or axis, in contrast to linear momentum, which is the inertia an object has due to movement along a straight line. The total angular momentum of an object or a system, such as the solar system, remains the same unless affected by something external. Angular momentum can be transferred internally between different parts of an isolated system, but the total remains the same.

archaea
One of three major groups of organisms on Earth. The archaea are similar in appearance to bacteria but are very different genetically.

Archean
The period in Earth's history between the end of the Hadean and 2.5 billion years ago.

asteroid family
A group of asteroids moving on similar orbits that are fragments from the catastrophic breakup of a larger body.

astronomical unit (AU)
Historically, the average distance between Earth and the Sun, but now defined more formally. Its value is 149,597,870 km (92,955,730 miles).

atomic mass
The mass of an atom of a specific isotope of an element, measured in units of one-twelfth of the mass of an atom of carbon-12.

atomic nucleus
The central, dense portion of an atom containing the great majority of its mass. The nucleus is composed of protons and neutrons.

atomic number
The number of protons in the nucleus of an atom. Atomic numbers are whole numbers and different for every chemical element.

atomic weight
The ratio of the average mass per atom of a sample of an element to one-twelfth of the mass of an atom of carbon-12. Different samples of the same element can have different atomic weights because the proportions of the various isotopes of the element are not the same. However, tables of standard atomic masses are published, based on average isotopic compositions. Atomic weight is also known as *relative atomic mass.*

basalt
A volcanic rock consisting mainly of the silicate minerals pyroxene and plagioclase.

basin
See impact basin

Big Bang
A theory for the origin and evolution of the universe according to which the universe began at a specific time from an infinitely compact state and has been expanding ever since. The cosmic background radiation that fills the universe is generally accepted to be evidence for the Big Bang model, as well as the observed expansion.

breccia
A rock made up of broken fragments cemented together by a finer-grained material. Breccias are a common outcome of impact processes.

calcium-aluminum-rich inclusions (CAIs)
Whitish, irregularly shaped particles found in chondritic meteorites. CAIs are typically 1 mm to a few cm (a fraction of an inch) in size, and are composed of rocky minerals with very high melting temperatures.

carbon-silicon cycle
The exchange of carbon between Earth's atmosphere and interior due to

geological processes. This process is believed to stabilize Earth's climate on long timescales, keeping the average temperature between the freezing and boiling point of water.

Centaur

A minor planet that moves largely between the orbits of Jupiter and Neptune.

chondrite

A common type of stony meteorite characterized by the presence of small rocky spheres, called chondrules. About 85 percent of meteorites are chondrites. A small number of meteorites are classed as chondrites even though they have no chondrules because they are otherwise so similar to chondrites generally.

chondrules

Small spheres of rock found in many stony meteorites. They range in size from less than 1 mm to more than 10 mm (0.025 to 0.25 inches) across and are made of silicate minerals that cooled very rapidly.

circumplanetary disk

A disk of material in orbit around a planet from which regular satellites can form.

cometary nucleus

A solid object composed of dust and ices that contains the bulk of a comet's mass. A comet's coma and tails form when ices in the nucleus evaporate. The resulting gas escapes into space carrying dust from the surface of the nucleus with it.

core

The dense, central portion of a body that has differentiated. The cores of the terrestrial planets are mostly composed of iron, and contain the bulk of each planet's siderophile elements.

core accretion model

A scenario in which a planetary embryo becomes massive enough for its gravity to capture gas from the surrounding protoplanetary disk, forming a gas giant planet.

crust

The outermost layer of a solid, differentiated planet.

C-type asteroid

One of several classes of asteroid identified on the basis of its spectrum. C-types tend to be dark, and some have features in their spectra showing that they contain hydrated clay-like minerals.

debris disk

A disk-shaped cloud of dust surrounding a star. A protoplanetary disk is believed to lose its gas and evolve into a debris disk when a star is a few million years old.

deuterium

An isotope of hydrogen, often called "heavy hydrogen" because its nucleus contains a neutron as well as a proton.

deuteron

The combination of one proton and one neutron, which forms the nucleus of deuterium.

differentiated

A planet or asteroid is differentiated if it has separated into layers of different density.

disk instability model

A scenario in which portions of a protoplanetary disk collapse due to gravity forming gaseous clumps that ultimately shrink to become gas giant planets.

DNA

Abbreviation for deoxyribonucleic acid, a large complex molecule that, in the form of a two-stranded helix, carries the genetic information of all living organisms (apart from some viruses).

dwarf planet

A small solar system body that (a) is in orbit around the Sun, (b) has sufficient mass for its self-gravity to overcome rigid body forces so that it assumes a hydrostatic equilibrium (nearly round) shape, (c) has not cleared the neighborhood around its orbit, and (d) is not a satellite. Examples include Ceres and Pluto.

electron

A negatively charged elementary particle. In normal atoms, a cloud of electrons surrounds the positively charged nucleus and the atom as a whole has no net charge. The way the electrons are arranged in an atom determines the atom's chemical properties. Electrons are also emitted when some radioactive nuclei decay. As a result of this kind of decay, a nucleus has one more proton and one less neutron.

emission line

In a spectrum, a narrow range of wavelength or frequency where there is a peak of intensity.

eukarya

Organisms with cells that have complex internal structures separated by membranes. Eukarya are one of the three major groups of organisms on Earth, along with bacteria and archaea. Examples include plants, animals, and fungi.

extremophile

An organism that is adapted to live in harsh conditions such as an environment with extreme temperatures or high salinity.

feeding zone

An annulus-shaped region surrounding the orbit of a planetary embryo.

The embryo grows mainly by sweeping up planetesimals orbiting within its feeding zone.

fusion
The merger of small atomic nuclei to form larger ones, typically releasing energy as a result.

gas giant
A planet mainly composed of hydrogen and helium, possibly containing a core made of denser materials. Jupiter and Saturn are examples.

Grand Tack model
A scenario in which Jupiter migrated inward and then outward across the asteroid belt while the terrestrial planets were growing. During this migration, the gravitational pull of Jupiter would have displaced most of the solid mass from the asteroid belt and the region that now contains Mars, explaining the low masses of each today.

greenhouse effect
The warming of a planet's surface due to the presence of greenhouse gases in its atmosphere, which trap outgoing infrared radiation. Common greenhouse gases are water vapor, carbon dioxide, and methane.

habitable zone
The region of space around a star where the temperature is just right for a solid, geologically active planet to have liquid water on its surface, if the planet also has an atmosphere providing enough surface pressure. The concept is based on the conditions required by life on Earth.

Hadean
The period in Earth's history between its formation and the end of the late heavy bombardment roughly 3.8 billion years ago.

helioseismology
The observation and study of how waves similar to sound waves propagate through the Sun. Helioseismology provides a way of probing the composition of the Sun's interior and the physical conditions there.

hit-and-run collision
A collision between two solid bodies at an oblique angle in which the objects slide past each other and separate again without gaining or losing much material.

hot spot
A region on the surface of a rocky planet where hot material from deep within the interior is welling up toward the surface. They are usually associated with volcanic activity.

hydrothermal vent
A fissure on land or under the sea where hot water, heated by geothermal energy, emerges from beneath the surface. Typically hydrothermal vents are found in volcanically active areas. In the ocean, where the pressure of overlying water is high, the water temperature can be over 400°C (750°F).

ice giant
A planet mainly composed of materials that form ices at low temperatures, such as water, methane, and ammonia. Uranus and Neptune are examples.

impact basin
A very large, shallow, circular depression on the surface of a planetary body caused by an impact.

impact melt
Glass-like material formed when an impact partially melts rocks in the target body.

infrared excess
More infrared radiation coming from the direction of a star than can be accounted for by the star itself. The extra radiation comes from dust grains close to the star that absorb visible starlight and give off infrared radiation.

irregular satellite
A moon following an orbit that is retrograde, highly inclined, or very elliptical.

isotopes
Variants of a particular chemical element that have different numbers of neutrons in their nuclei and therefore have different atomic masses. Most elements have more than one stable isotope. For example, oxygen has three naturally occurring isotopes, oxygen-16, oxygen-17, and oxygen-18. The nuclei of all of them contain 8 protons, defining the element and its chemical properties, but the three different isotopes contain 8, 9, and 10 neutrons, respectively.

Jupiter-family comet
One of several hundred known comets that have orbital periods of less than 200 years and orbital planes tilted only slightly with respect to the plane containing the planets.

K, kelvin
The Kelvin scale (named after the physicist Lord Kelvin) is a system of temperature measurement widely used in science. Its zero point is absolute zero. Temperatures and temperature differences are measured in kelvins (written without a capital letter), abbreviated to K. A kelvin is the same as 1 degree on the Celsius scale.

Kirkwood gaps
Narrow ranges of orbits in the asteroid belt in which there are very few asteroids. The gaps occur where an asteroid's orbit would be in a resonance with Jupiter's orbit—that is, its orbital period would be in a whole-number ratio with Jupiter's orbital period (e.g., 3:1, 5:2, etc.).

Kuiper belt
A belt of icy minor planets located beyond the orbit of Neptune and extending to roughly 50 AU from the Sun.

late heavy bombardment
An episode roughly 3.9 billion years ago in which many of the impact basins on the Moon formed. Each of the terrestrial planets probably experienced many large impacts at the same time.

late veneer
The last component of material added to a growing terrestrial planet after its core has finished forming. Earth's late veneer supplied most of the highly siderophile elements present in the crust today.

lava
Hot, molten rock that has erupted onto the surface of a planet.

light-year
The distance traveled by light moving through a vacuum in one year. Roughly 10 trillion km or 6 trillion miles.

lipid
A fatty polymer that forms biological membranes including cell walls.

lithophile element
An element that preferentially enters a planet's rocky mantle and crust rather than its core when a planet melts and differentiates. Examples include sodium, magnesium, aluminum, and hafnium.

lithosphere
The rigid outermost shell of a rocky planet such as Earth. On Earth, the lithosphere consists of the crust and the top layer of the mantle. Together they slowly move over the less rigid layers of mantle below the lithosphere.

long-period comet
A comet with an orbital period longer than 200 years. Long-period comets can have orbits tilted at any angle with respect to the plane containing the planets.

magma
Hot, molten rock beneath the surface of a planet.

magma ocean
A temporary layer of liquid rock formed in the outer portion of a rocky planet by a large impact.

magnitude
A scale used to measure the brightness of stars and other astronomical objects. The higher the magnitude, the fainter the object. Sirius, the brightest star in the sky has a magnitude of −1.5, while the faintest stars visible to the naked eye are about magnitude +6.

mantle
The portion of a solid planet that lies above its central core. On terrestrial planets, the mantle is largely composed of rocky silicates.

mare (plural maria)
The Latin word for "sea," which is used for dark areas on the Moon. These

areas are plains of basaltic rock that formed when lava erupted and filled large impact basins.

matrix
Fine-grained, compacted dust that fills the space between chondrules in a chondritic meteorite.

metallicity
A measure of the abundance of elements heavier than helium in a star.

meteor
The luminous trail seen when a dust particle or larger piece of rock from space burns up in Earth's atmosphere.

meteorite
A rock that has traveled through space, then landed on the surface of Earth or any other body in the solar system.

meteoroid
A piece of rock or dust in space that could become a meteor or a meteorite. This term is usually applied to objects smaller than about 100 meters (300 feet), while larger objects are more often described as asteroids.

microgravity
A situation, such as in space, in which people and things experience weightlessness. Weightlessness is not caused by the absence of gravity but is the result of things moving or falling together freely through space, as happens when they are in orbit around Earth or the Sun, for example.

minor planet
A body orbiting the Sun that is smaller than a major planet and that doesn't have a coma or tail like a comet. Minor planets orbiting closer to the Sun than Jupiter are commonly called asteroids.

molecular cloud
A large, tenuous cloud of gas and dust in space containing up to a million solar masses of material. Molecular clouds are often the sites of new star formation.

molecular cloud core
A small, dense portion of a molecular cloud that can collapse due to gravity to form a star.

nebular hypothesis
The idea that the solar system originated from a disk-shaped cloud of material surrounding the Sun. It was originally developed by Immanuel Kant and Pierre Simon Laplace.

neutrino
An elementary particle that rarely interacts with other forms of matter. Neutrinos are produced in large numbers by nuclear reactions inside stars.

neutron
An elementary particle with no electric charge that is found in atomic

nuclei. The atomic nuclei of all elements apart from hydrogen contain neutrons.

neutron star

A star that has collapsed so much under its own weight that its electrons and protons have merged to make neutrons. A typical neutron star is only about 10 km (6 miles) across but has between 1.5 and 3 times the mass of the Sun.

Nice model

A scenario for the early evolution of the solar system in which the giant planets formed closer together than they are today and migrated to their current locations. During this migration, the giant planets temporarily acquired highly elongated orbits, displacing many asteroids and comets due to their gravity. It is named after the city of Nice in France where the model was developed.

noble gases

A group of chemical elements that exist as gases at normal temperatures and pressures on Earth, and form very few chemical compounds with other elements. The noble gases are helium, neon, argon, krypton, and xenon, which are stable, and radon, which is radioactive.

obliquity

The angle between a planet's equator and the plane containing its orbit. Planets with an obliquity near 90 degrees rotate on their sides. Planets with obliquities greater than 90 degrees have retrograde rotations.

oligarchic growth

A stage in the growth of planets from planetesimals. During oligarchic growth, each region of a protoplanetary disk is dominated by a single planetary embryo that grows by sweeping up planetesimals in its vicinity.

Oort cloud

A swarm of billions of dormant comets orbiting the Sun tens of thousands of AU beyond the planets. Occasionally, the gravitational tug from a passing star or the Milky Way pulls an object out of the Oort cloud sending it close to the Sun where it becomes a visible comet.

perihelion

The point where an object orbiting the Sun is closest to the Sun.

photoevaporation

The heating and acceleration of gas as it absorbs ultraviolet light, allowing the gas to escape from an object's gravity. Photoevaporation is one way in which gas can be removed from a star's protoplanetary disk.

photon

A particle of electromagnetic radiation, such as light. Electromagnetic radiation has both wave-like and particle-like properties, and a photon is like a tiny "packet" of waves.

phylogenetic
Relating to the study of the evolutionary relationship between living organisms.

planet
An astronomical body that is not massive enough to become a star but not so small that it is classified as an asteroid or comet. The upper mass limit for a planet is about 13 times the mass of Jupiter. In the solar system, the eight largest planets are called the major planets, and astronomers also define a class of smaller dwarf planets.

planetary embryo
An object intermediate in mass between a planetesimal and a fully grown planet that exists during the oligarchic growth stage of planet formation.

planetesimal
A small, solid body in a protoplanetary disk that forms the basic building block of planets.

plate tectonics
The theory describing the evolution of Earth's lithosphere over millions of years. The lithosphere is divided into roughly a dozen continent-sized plates that move slowly with respect to one another due to motions in the underlying mantle. The motion of plates causes earthquakes, volcanic activity, and the formation and destruction of oceanic crust.

Plutino
A member of the Kuiper belt that lies in an orbital resonance with Neptune. Plutinos are named after the most prominent example, Pluto.

primitive body
A solid object that never grew hot enough to melt and differentiate.

proplyd
A contraction of "protoplanetary disk."

proton
A positively charged elementary particle found in atomic nuclei. The nucleus of a hydrogen atom is a single proton. The number of protons in the nucleus of an atom is called the atomic number and defines which element it is.

protoplanetary disk
A disk-shaped cloud of gas and dust surrounding a young star that is the site where a planetary system forms.

radioactivity
The emission of particles or radiation when the unstable atomic nuclei in radioactive materials decay. Three different kinds of emissions from radioactive materials were labeled alpha, beta, and gamma by early researchers. In alpha decay, a nucleus emits an "alpha particle," which consists of two protons and two neutrons and is identical to the nucleus of a helium atom.

In beta decay, a nucleus emits an electron. Some nuclei also emit gamma rays, a powerful form of electromagnetic radiation, as part of the decay process.

radiometric dating
One of several techniques used to measure the age of an object using the radioactive decay of materials trapped inside it.

red giant
A star that is in a late stage of its evolution and has expanded greatly in size.

refractory
Describing a material with a high melting temperature.

regolith
A layer of loose, fine-grained material on the surface of the Moon or any planetary body.

regular satellite
A natural satellite of a planet that travels along a direct (i.e., not retrograde) orbit, which is in the planet's equatorial plane and is nearly circular rather than markedly elliptical.

resonance
A situation in which one orbiting body, such as a planet or asteroid orbiting the Sun, is subject to a systematic gravitational disturbance by another orbiting body at regular intervals. Resonances occur when the orbital periods of the two bodies concerned are in a whole-number ratio, such as 2:1.

retrograde
Describing orbital motion of rotation that is in the opposite direction to the general direction of movement of most objects in the solar system. In an overview from above (i.e., north of) the plane of the solar system, retrograde motion is clockwise.

RNA
Abbreviation for ribonucleic acid, a complex molecule essential to life. Like DNA, RNA can carry genetic information, and many viruses use RNA rather than DNA.

Roche limit
The closest distance to a planet that a satellite held together solely by gravity can orbit without being torn apart by tidal forces. If the planet and the satellite have the same density, the Roche limit, measured from the center of the planet, is about 2.5 times the planet's radius. Solid satellites can orbit inside the Roche limit if they are strong enough.

rubble pile
An asteroid or comet composed of several pieces held together solely by gravity.

scattered disk
A group of minor planets orbiting beyond Neptune that can have close

encounters with Neptune. These objects typically have tilted and elongated orbits as a result.

sedimentary rock

A rock formed when sand, clay, and silt settle to the bottom of a body of water and become compressed by the weight of additional layers of material.

seismometer

An instrument used to measure waves passing through Earth generated by earthquakes.

siderophile element

An element with a chemical affinity for iron. Siderophile elements preferentially enter a planet's core when a planet melts and differentiates. Examples include gold, platinum, nickel, and tungsten.

silicate

A chemical compound containing the element silicon, which in most silicates is bound with oxygen. Silicate minerals are the major constituent of rocks in Earth's crust and elsewhere in the solar system.

snowball Earth

One of several episodes in the distant past when Earth was almost entirely covered in ice.

solar nebula

The protoplanetary disk that surrounded the Sun shortly after it formed. The solar nebula is believed to be the site where the planets and other members of the solar system formed.

space weathering

The change in the appearance of the surface layer of a rocky object in space caused by radiation from the Sun, cosmic rays, and impacts.

spiral arm

A region of higher than average density in the disk of a spiral galaxy. Spiral galaxies typically have two or more arms winding outward from a central bulge or bar. They are marked by bright, young stars embedded in glowing gas clouds, and molecular clouds.

spiral nebula

An obsolete term for a spiral galaxy. The nature of spiral nebulae remained unclear until the early 20th century, when astronomers realized these objects were other galaxies outside the Milky Way.

S-type asteroid

One of several classes of asteroid identified on the basis of its spectrum. S-types tend to be bright and have spectral features that suggest they are composed of silicate rocks.

stagnant lid

The behavior of the outermost layer of a rocky planet that doesn't undergo plate tectonics. The crust of such a planet forms a single static plate rather than the many mobile plates found on Earth. Examples include Mars and Venus.

supernova

A catastrophic explosion of a star that generates tremendous amounts of energy and can be seen from a great distance. A supernova occurs either when a massive star at the end of its life runs out of nuclear fuel and implodes, or when a white dwarf in a binary star system captures enough material from its companion to trigger runaway nuclear fusion. Nuclear reactions that occur during supernova explosions are an important source of heavy elements.

synchronous rotation

The situation in which the rotation and orbital periods of a planetary satellite are the same so that the satellite always keeps the same face toward its parent planet.

tidal forces

The stretching or distortion of a body, such as a moon or planet, that results when different parts of it experience gravitational forces of unequal strength. For example, Earth experiences a tidal force caused by the Moon because the Moon pulls more strongly on the side of Earth nearest to it than on the farther side.

transition disk

A disk of gas and dust surrounding a young star that is evolving from a protoplanetary disk to a debris disk.

trans-Neptunian object (TNO)

A minor planet that orbits the Sun beyond Neptune. TNOs include members of the Kuiper belt and the scattered disk.

tree of life

A diagram showing the genetic relationships between different groups of organisms. Typically, closely related species are located close to one another on neighboring branches of the tree.

Trojan

An object that shares an orbit with a planet, usually in a location that is about 60 degrees ahead or behind the planet as seen from the Sun.

T Tauri star

A young star that derives energy from its slow contraction rather than the fusion of hydrogen into helium in its interior. T Tauri stars, which are named after a prominent example in the constellation Taurus, are often accompanied by a protoplanetary disk.

uniformitarianism

The view that Earth has changed gradually and continually over time due to the same geological processes that operate today.

volatile

Describing a material with a low melting temperature.

white dwarf

A dying star that has exhausted all its sources of nuclear energy and has collapsed under its own weight until the atomic nuclei and electrons it contains are packed tightly together. A white dwarf is what remains after a red giant star has blown off its outer layers.

Yarkovsky effect

The gradual acceleration of a small body caused by absorbing sunlight and emitting infrared radiation in slightly different directions due to the body's rotation. The orbits of small asteroids change significantly over millions of years due to the Yarkovsky effect.

zircon

A mineral found in granite and other continental rocks. Zircons can survive when their parent rocks are destroyed. They are particularly useful for radiometric dating.

SOURCES AND FURTHER READING

This is a selected list of some of the books and articles that we used in writing this book, including some of historical interest, together with a variety of recommended further reading and reference resources relating to planetary sciences.

Adams, Fred C. "The Birth Environment of the Solar System." *Annual Reviews of Astronomy and Astrophysics*, vol. 48:47–85, 2010.

Asplund, Martin, Grevasse, Nicolas, Sauval, A. Jacques, & Scott, Pat. "The Chemical Composition of the Sun." *Annual Reviews of Astronomy and Astrophysics*, vol. 47:481–522, 2009.

Barucci, Maria A., Boehnhardt, Hermann, Cruikshank, Dale P., & Morbidelli, Alessandro (eds.). *The Solar System Beyond Neptune*. University of Arizona Press, 2008.

Beatty, J. Kelly, Petersen, Carolyn C., & Chaikin, Andrew (eds.). *The New Solar System* (4th ed.). Sky Publishing and Cambridge University Press, 1999.

Bergin, Edwin A., & Tafalla, Mario. "Cold Dark Clouds: The Initial Conditions for Star Formation." *Annual Reviews of Astronomy and Astrophysics*, vol. 45:339–96, 2007.

Boss, Alan. *The Crowded Universe: The Race to Find Life Beyond Earth*. Basic Books, 2009.

Brush, Stephen G. *A History of Modern Planetary Physics* (3 vols.: 1. *Nebulous Earth*, 2. *Transmuted Past*, 3. *Fruitful Encounters*). Cambridge University Press, 1996.

Canup, Robin M. "Simulations of a Late Lunar-Forming Impact." *Icarus*, vol. 168:433–68, 2004.

Dalrymple, G. Brent. *Ancient Earth, Ancient Skies*. Stanford University Press, 2004.

de Pater, Imke, & Lissauer, Jack J. *Planetary Sciences* (2nd ed.). Cambridge University Press, 2010.

Fernández, Julio A. "On the Existence of a Comet Belt Beyond Neptune." *Monthly Notices of the Royal Astronomical Society*, vol. 192:481–91, 1980.

Holmes, Arthur. *The Age of the Earth*. Harper & Brothers, 1913.

Jackson, Patrick Wyse. *The Chronologers' Quest: The Search for the Age of the Earth*. Cambridge University Press, 2006.

Jaki, Stanley L. *Planets and Planetarians*. Wiley, 1978.

Jet Propulsion Laboratory. *Solar System Dynamics* (website). http://ssd.jpl.nasa.gov/ (includes a wide range of regularly updated solar system information, listed on the site map).

Jewitt, David. "The Discovery of the Kuiper Belt." *Astronomy Beat: Astronomical Society of the Pacific* (online newsletter for members), no. 48, 2010.

Kasting, James. *How to Find a Habitable Planet*. Princeton University Press, 2009.

Knell, Simon J., & Lewis, Cherry L. E. "Celebrating the Age of the Earth." *Geological Society, London, Special Publications*, vol. 190:1–14, 2001.

Kowal, Charles T. *Asteroids: Their Nature and Utilization*. Ellis Horwood, 1988.

Kuiper, Gerard P. "On the Origin of the Solar System." *Proceedings of the National Academy of Sciences*, vol. 37:1–14, 1950.

Lang, Kenneth R., & Gingerich, Owen (eds.). *A Source Book in Astronomy and Astrophysics 1900–1975*. Harvard University Press, 1979.

Lewis, Cherry. *The Dating Game: One Man's Search for the Age of the Earth*. Cambridge University Press, 2000.

Lunine, Jonathan I. *Earth: Evolution of a Habitable World* (2nd ed.). Cambridge University Press, 2008.

McFadden, Lucy-Ann, Weissman, Paul, & Johnson, Torrence (eds.). *Encyclopedia of the Solar System* (2nd ed.). Academic Press, 2006.

McSween, Harry Y., Jr. *Meteorites and Their Parent Planets*. Cambridge University Press, 1999.

Norton, O. Richard. *The Cambridge Encyclopedia of Meteorites*. Cambridge University Press, 2002.

Patterson, C. "Age of Meteorites and the Earth." *Geochimica et Cosmochimica Acta*, vol. 10:230–37, 1956.

Russell, Henry Norris. *The Solar System and Its Origin*. Macmillan, 1935.

Taylor, Stuart Ross. *Solar System Evolution: A New Perspective* (2nd ed.). Cambridge University Press, 2001.

Whipple, Fred L. *The Mystery of Comets*. Smithsonian Institution, 1985.

Williams, J. P. "The Astrophysical Environment of the Solar Birthplace." *Contemporary Physics*, vol. 51:381–96, 2010.

Woolfson, Michael. *The Formation of the Solar System: Theories Old and New*. Imperial College Press, 2007.

1992 QB1, 248, 250
1999 RQ36, 271
2008 TC3, 84
51 Pegasi, 15

achondrites, 82–83
Adams, John Couch, 38, 39
Airy, George Biddell, 38
Alfvén, Hannes, 54
alpha elements, 95
alpha particles, 95, 97–98
alpha radiation, 62
Andromeda galaxy, 52
angular momentum, 49; angular momentum problem, 50, 51, 55; of Earth-Moon system, 172; of solar system, 3, 48, 54
Apollo missions, 86, 169, 177, 183
archaea, 192
Archean era, 191, 197, 199
Aristarchus of Samos, 23
Aristotle, 23, 24, 28, 29
asteroid belt, 7, 36, 123, 225–26, 228–31, 234–36, 266, 273
asteroid families, 231–33
asteroids, 7, 8, 35, 36, 42; collisions between, 77, 79, 90, 226–29, 265, 273–74; C-type, 85, 229, 237; D-type, 229, 261; as meteorite parent bodies, 76, 77, 83–85, 150, 241; near-Earth, 79; orbits of, 229, 232; P-type, 229; satellites of, 236; shapes of, 236; space missions to, 236–41; S-type, 85, 229, 239; Trojan, 7, 42, 259, 260, 261. *See also* asteroid belt; asteroid families
Astraea, 36
astronomical unit, 5, 19, 20
atomic mass, 65, 92, 93
atomic number, 65, 93
atomic weight, 92, 93, 94
atoms: first, 98; structure of, 64

Babinet, Jacques, 49
Babylonian astronomy, 22
bacteria, 192
Baptistina family of asteroids, 233
Becquerel, Henri, 61
Beta Pictoris, 125, 126, 127
beta radiation, 62
Big Bang, 73, 96, 97
Biot, Jean-Baptiste, 37
black hole, 105
Bode, Johann Elert, 32
Bode's law, 32–33, 34
Boltwood, Bertram, 63
Boss, Alan, 210
Bottke, William, 235
Brahe, Tycho, 28
Brown, Michael, 251
Buffon, Comte de, 45
Burbidge, Margaret and Geoffrey, 99, 100

calcium-aluminum-rich inclusions, 82, 83, 89, 90, 131
Callisto, 217, 218, 219, 220
Cameron, Alastair, 178
Canup, Robin, 224
capture hypothesis (of Moon's origin), 175–76
carbonaceous chondrites, 82, 85, 90, 152
carbon-nitrogen-oxygen (CNO) cycle, 100–102
carbon-silicon cycle, 200, 201, 204
Cassini mission, 41
Cassini, Jean-Dominique, 20–21

Centaurs, 246–47, 250–51
Ceres, 6, 7, 270; composition of, 241; discovery of, 34–35; size of, 225; spectrum of, 241
Chaldni, Ernst, 37
Challis, James, 38
Chamberlin, Thomas Chrowder, 51, 53, 54, 61
Chamberlin-Moulton theory, 53, 54
Charon, 40, 253, 256, 257, 265, 270
Chiron, 246–47
chondrites, 81–82, 89, 133; composition of, 95; parent bodies of, 85, 91, 239
chondrules, 81–83, 89, 90, 91, 130–31, 133–34, 137, 142, 265
Churyumov-Gerasimenko, comet, 17, 269
CIAs, 82, 83, 89, 90, 131
Clementine spacecraft, 169
CNO cycle, 100–102
coaccretion hypothesis (of Moon's origin), 176
comets, 7, 17, 17, 242; composition of, 255–56; formation of, 244; Jupiter-family, 8, 123, 242, 247, 254; long-period, 8, 123, 242, 243, 247; orbits of, 30, 31; water content of, 152. *See also* Oort cloud
Cook, James, 19
Copernicus, Nicolaus, 25–27, 29
core accretion model, 210, 211–14
Cosmic Background Explorer (COBE) mission, 73
cosmic microwave background, 73
cosmic rays, 77
craters: on Mars, 11, 12, 163–64, 184; on Mars's moons, 163; on Mercury, 156, 184; on the Moon, 168–70, 171, 183–84, 185; on Venus, 160
cubewanos, 249
Curie, Marie and Pierre, 61
cyanobacteria, 197

Dactyl, 236, 237
Darwin, Charles, 43–44
Darwin, Erasmus, 44
Darwin, George, 58, 174–75
Davis, Donald, 178
Dawn mission, 228, 241, 270
debris disks, 127–28, 137, 139
Deep Impact mission, 17, 225
Deimos, 163
detached objects, 251
disk instability model, 210, 214–15
DNA, 191, 193, 194, 196
Dunthorne, Richard, 173–74
dust: in chondrules, 133–34; in circumstellar disk, 124, 125, 126–27, 129, 264; coagulation of, 132–33, 138, 265; in comets, 244, 255–56; in planetary envelopes, 212; in planetesimals, 142; in solar nebula, 131–32, 133, 135
dwarf planet, 7
dynamical friction, 143, 144, 146
dynamo effect, 151
Dysnomia, 250

Eagle nebula, 113–14
Earth: age of, 56–69; atmosphere of, 186, 197–98, 199; climate of, 200–203; compared to Venus, 140–41; composition of, 95; core of, 149, 150, 151, 187; crust of, 149, 150, 187–89; early evolution of, 186–91, 265; formation time of, 147; habitability of, 204, 272; impacts on, 187; life on, 191–200, 268; lithosphere of, 152, 153; magnetic field of, 151; mantle of, 149, 150, 152, 153, 187; obliquity of, 182; plate tectonics on, 152, 153, 188–89, 200, 201; rotation of, 182; size of, 21; structure of, 148–150; water on, 151, 152, 153, 186, 188, 195, 199, 265. *See also* greenhouse effect; late heavy bombardment
Eddington, Arthur, 71
Edgeworth, Kenneth, 244, 248, 250
EGGS, 114
elements, chemical: lithophile, 150; periodic table of, 92–94; siderophile, 149–50; solar system abundance of, 95–96
Elst-Pizzaro, 8, 9
Enceladus, 41
encounter theory, 53, 54
enstatite chondrites, 82, 241
epicycles, 24, 26
Epsilon Eridani, 125
equant, 24, 26
Eris, 6, 7, 250
Eros, 7, 41, 236–39
Eudoxus of Cnidus, 23
eukarya, 193, 198
Europa, 217, 218, 219, 220
evaporating gaseous globules (EGGS), 114
extrasolar planets: discovery of, 15; migration of, 267; search for, 272–73
extremophiles, 193

Fernández, Julio, 247, 248
Ferrell, William, 174
Fischer, Osmond, 174
fission theory (of Moon's origin), 174–75
Flamsteed, John, 30
Fomalhaut, 125

Fowler, Willy, 99, 100
Frail, Dale, 15
Frost, Edwin B., 50

Galilean moons, 217, 218, 219
Galilei, Galileo, 28–29, 39, 221
Galileo mission, 41, 217, 236
Galileo probe, 206–7
Galle, Johann Gottfried, 38–39
gamma rays, 62, 99
Ganymede, 217, 218, 219, 220
gas giants, 209–10. *See also* giant planets; Jupiter; Saturn
Gaspra, 236
Gauss, Karl Friedrich, 35
Gefion family of asteroids, 232–33
Gerling, E. K., 67–68
giant impact, 147; as Charon's origin, 256; on Mercury, 154–55; as Moon's fate, 182; as Moon's origin, 177–81, 186; on Uranus, 216
giant planets, 5; atmospheres of, 208; circum-planetary disks of, 219; composition of, 210; formation of, 210–15, 259, 266, 271; migration of, 162, 258, 260, 262; natural satellites (moons) of, 5, 216–21; obliquity of, 216; physical properties of, 208–9; rings of, 222–24; rotation of, 215–16; space missions to, 41; structure of, 207–8. *See also* Jupiter; Neptune; Saturn; Uranus
giant star. *See* red giant star
Giotto mission, 41
globular cluster, 109
Goldreich, Peter, 136
Gomes, Rodney, 259
Grand Tack theory, 162, 231
gravitational focusing, 143, 144, 146
gravitational instability, 136, 215
gravity: Einstein's theory of, 64; of Jupiter, 8, 31, 79, 152, 159, 162, 213, 217, 229, 231, 260; of Mars, 166, 201, 265; of Mercury, 156, 265; of the Moon, 172, 174; of Neptune, 249, 251, 258; Newton's theory of, 29, 30, 31; and planet formation, 136–37, 139, 142–43, 145, 146, 210–13; of Pluto, 40; of Saturn, 152, 214, 224, 258; and star formation, 113, 115; and stellar structure, 102–3, 105; of Sun, 3, 45, 129, 134, 135, 155, 174, 213, 243, 266; of Uranus, 258
greenhouse effect, 200; on Earth, 196, 199, 200, 202, 203; on Mars, 166; on Venus, 158
Gregory, David, 32
Gregory, James, 21

habitable zone, 200–201
Hadean era, 186–91
half-life, 14, 62, 65
Halley, comet, 31, 41
Halley, Edmond, 21, 31, 58, 173
Hansen, Brad, 161
Harding, Karl, 35
Haro, Guillermo, 116
Hartley 2, comet, 17, 18
Hartmann, William, 178
Haumea, 6
Hebe, 36
heliocentric theory, 26
helioseismology, 72, 73, 264
Helmholtz, Hermann von, 47, 59
Hencke, Karl Ludwig, 36
Herbig, George, 116
Herbig-Haro objects, 116
Herschel, William, 33, 35, 48
Hester, Jeff, 113
Hidalgo, 7
Hirayama, Kiyotsugu, 231–32
Holmes, Arthur, 67–68
Houtermans, Fritz G., 67–68
Hoyle, Fred, 54, 99, 100, 103
Hubble, Edwin, 63
Hubble's law, 64
Huygens, Christiaan, 222
Hyabusa mission, 42, 84, 239
Hyabusa 2 mission, 271
Hydra, 253

ice giants, 210, 213. *See also* giant planets; Neptune; Uranus
Ida, 236, 237
inertia, rotational. *See* angular momentum
Infrared Astronomical Satellite (IRAS), 124, 125
infrared excess, 124–25, 128
interstellar medium, 109, 111; enrichment of, 107, 121–22
Io, 41, 217, 218, 220
irregular satellites, 217–18, 261; collisions of, 221; origin of 220–21, 259, 260, 265
isochron, 66, 67, 69
isotopes, 64, 66, 77, 93; radioactive, 65, 67, 70, 90, 91, 120, 150, 265
Itokawa, 84, 85, 239

Jeans, James, 53
Jeffreys, Harold, 53, 175
jets, in star-forming regions, 115, 117
Jewitt, David, 248

Joly, John, 58
Juno (asteroid), 35
Juno mission, 271
Jupiter: atmosphere of, 205, 206–7; circumplanetary disk of, 219; core of, 205–6, 207, 213, 271; interior of, 205–8; magnetic field of, 207; moons of, 217–18; rings of, 222; space missions to, 11, 271. *See also* giant planets
Jupiter-family comet. *See* comet

Kant, Immanuel, 46–47, 123, 174
Karin family of asteroids, 232
Karnak, 1
Kasting, James, 200
Kelvin, Lord (William Thomson), 59–60
Kepler mission, 15–16, 272–73
Kepler, Johannes, 27–28, 29, 44
Kepler's laws, 28, 30
King, Clarence, 60
Kirkwood, Daniel, 50, 77
Kirkwood gaps, 77–78
Koronis family of asteroids, 232
Kowal, Charles, 6, 246
Kuiper belt, 7–8, 42, 119, 121, 123, 256, 259, 261, 262, 265, 266, 270; classical, 249, 250, 259; future evolution of, 273–74; resonances in, 250
Kuiper, Gerard, 244, 248

L'Aigle meteorite fall, 37
Laplace, Pierre Simon de, 16, 44–45, 46, 47, 123, 176
Laskar, Jacques, 274
Lassell, William, 39
last common ancestor, 193
late heavy bombardment: of asteroids, 228; of Earth, 190; of Mars, 228; of Mercury, 228; of Moon, 183–85; and planetary migration, 259, 268
Leclerc, Georges-Louis, Comte de Buffon, 45
Leighton, Robert, 72
Leverrier, Urbain Jean Joseph, 38, 39
Levison, Harold 259
Lichtenberg, Christopher, 37
LINEAR survey, 77
lithophile elements, 150
long-period comet. *See* comet
Lowell, Percival, 10, 39
Luna spacecraft, 169
Lunar Prospector spacecraft, 169
lunar rocks, 10, 69, 70, 86, 169, 183
Lutetia, 240–41
Luu, Jane, 248
Lyell, Charles, 59

Magellan spacecraft, 159
Maillet, Benoit de, 57
main-sequence star, 72
Makemake, 6
Malhotra, Renu, 258
maria (seas), lunar, 169, 170, 172, 184
Mariner 4, 10–11, 12
Mars: atmosphere of, 166–67, 201, 265; composition of, 95; core of, 163; craters on, 11, 12, 163–64, 184; formation of, 70, 163; life on, 10, 11, 167; lithosphere of, 164; magnetic field of, 164; moons of, 163; obliquity of, 182; orbit of, 4, 28; polar caps of, 165; size of, 161; space missions to, 10–11, 41; surface of, 161, 163–65, 184; water on, 165, 166
Maskelyne, Nevil, 33
mass extinctions, 274
Mathilde, 41, 237–38
Maxwell, James Clerk, 50
Mayer, Robert, 174
Mayor, Michel, 15
Mendeleyev, Dimitri, 92, 93, 94
Mercury: atmosphere of, 154, 265; craters on 156, 184; formation of, 153–57; magnetic field of, 155–56; orbit of, 4, 155; physical properties of, 153, 154, 265; rotation of, 155, 156; structure of, 154, 156; surface of, 154, 156, 184
Merrill, Paul, 94, 104
Messenger spacecraft, 154, 157
meteorites, 37, 79, 184; and age of Earth, 56, 68–69; Antarctic, 87–88; classification of, 80–83, 86; falls of, 37, 75–76, 84; HED, 84, 90, 226; iron, 80–81, 85, 91, 150, 234, 235–36, 240; lunar, 86–87, 88, 169; Martian, 69, 70, 86–87, 163, 164; M-type, 85, 240; organic chemicals in, 195; parent bodies of, 76, 77, 83–87, 90, 150, 233–36; as radiometric clocks, 90; space weathering of, 84–85; stony, 81–83; water content of, 152. *See also* achondrites; chondrites
meteoroids, 76–77, 78
meteors, 76
meter-size barrier, 135, 136
methanogens, 196, 198
micrometeorites, 85
midocean ridges, 188

migration of planets, 16, 162, 214, 258, 260, 262
Milky Way galaxy, 108–11; halo of, 109; spiral arms of, 109; Sun's location in, 110
Miller, Stanley, 194
minor planets, 7. *See also* asteroids
missing mantle problem, 233–36
model, mathematical, 14–15
molecular clouds, 111, 113; cores in, 113, 114; Orion, 111–12, 115
Moon rocks, 10, 69, 70, 86, 169, 183
Moon: composition of, 95, 170–72; core of, 172; craters on, 168–70, 171, 183–84, 185; crust of, 172, 183; equatorial bulge of, 171; formation of, 58, 71, 174–81, 186, 265; impact basins on, 183, 185; mantle of, 172; orbit of, 172–74, 181–82, 274; rotation of, 174; space missions to, 11, 169; surface of, 169–70, 183–85; tidal forces on, 181–82; volcanism on, 184; water on, 171; *See also* late heavy bombardment; maria, lunar; meteorites, lunar; Moon rocks; moonquakes
moonquakes, 172
Morbidelli, Alessandro, 259
Moseley, Henry, 93, 94
Moulton, Forest Ray, 51, 53

NEAR Shoemaker mission, 41, 236–39
nebular hypothesis, 16, 45, 46, 47, 50, 51, 52, 54
Neptune: atmosphere of, 208; core of, 213; discovery of, 38–39; migration of, 258–59, 260; moons of, 218, 221; rings of, 222. *See also* giant planets
neutrinos, 104, 105, 106
neutron star, 105
New Horizons mission, 11, 253, 257, 269–70
Newton, Isaac, 29–31, 44, 46
Nice model, 259–62
Nix, 253
Nölke, Friedrich, 53
nucleosynthesis: in early universe, 96–98; in stars, 99–105

O'Dell, Robert, 117
Olbers, Heinrich, 35
oligarchic growth, 145, 146, 211
Olympus Mons, 161, 164
Oort cloud, 9, 120, 123, 214, 243, 247, 251, 258, 260, 274
Oort, Jan, 243
Öpik, Ernst, 79, 243
Orion nebula, 111, 112
OSIRIS Rex mission, 271

Pallas, 35, 225
Parsons, William, Earl of Rosse, 48, 49
Patterson, Clair C., 56, 68–69
Peekskill meteorite, 75–76
Perrier, Carlo, 94
Phobos, 50, 163, 274
Pholus, 247
photoevaporation, 113, 114, 138
photosynthesis, 196, 203
Piazzi, Guiseppe, 34
Picard, Jean-Felix, 21
Pickering, William Henry, 39
"pillars of creation," 113, 114
Planet X, 39, 40
planetary embryos, 144–47, 152, 161, 211–12; in asteroid belt, 230–31; collisions between, 146–47, 150; gas accretion onto, 212–13
planetesimals, 53, 54, 55, 136, 137, 141–47, 152, 177, 184, 214, 231, 260–61, 267; collisions between, 142–43; future orbits of, 274; growth of, 143, 230; heating of, 142
planets: migration of, 16, 162, 214, 258, 260, 262; naming of, 22; orbits of, 32; water content of, 151. *See also* extrasolar planets; giant planets; terrestrial planets
plate tectonics, 153, 159, 160, 188–89, 201; on Mars, 165, 166, 201
Plutinos, 249, 258
Pluto, 6, 7, 11, 42, 250, 256, 270; discovery of, 39–40; orbit of, 40, 249, 253, 258; moons of, 253; surface of, 253, 254
Poincaré, Henri, 46
proplyds, 117
proton-proton chain, 100, 101
protoplanetary disk, 117, 118, 128, 129, 137, 138–39, 264, 267
protostars, 115, 116
Ptolemy, Claudius, 21, 22, 24

Queloz, Didier, 15

Rabinowitz, David, 251
radioactivity, 13–14, 62, 65, 94; discovery of, 61
radiometric dating, 14, 62–63, 64–68, 70, 73, 74, 150, 228, 264, 265
red dwarf star, 102
red giant star, 72, 81, 103, 104, 106, 109, 121
regolith, lunar, 170

regular satellites, 217; formation of, 219–20; orbital resonances between, 220
resonances, orbital: asteroids in, 78, 84, 229–30; with giant planets, 230–31; between Jupiter and Saturn, 260; in the Kuiper belt, 250; meteoroids in, 78; with Neptune, 249, 250; regular satellites in, 220
retrograde motion, 23
Rheasilvia (impact basin), 228, 233
Rheticus, Georg Joachim, 26
Richer, Jean, 20–21
ring systems, 5. *See also* Jupiter; Neptune; Saturn; Uranus
Ringwood, Alfred, 175
RNA, 192, 193, 194, 195–96
Roche, Édouard, 176
Roche limit, 180
Rosetta mission, 17, 240–41, 269
Rosetta stone, 1–2
Rosse, Earl of, 48, 49
r-process, 106
Ruskol, Evgenia, 177
Russell, Henry Norris, 53, 54
Rutherford, Ernest, 61–62, 63

Safronov, Viktor, 55, 135–36
Saturn: moons of, 218; rings of, 31–32, 41, 221–24; space missions to, 11. *See also* giant planets
scattered disk, 250–51, 258, 261
Scheila, 77
Schmidt, Otto, 55
Sedgwick, Adam, 48
Sedna, 6, 119, 251–52
Segrè, Emilio, 94
seismometry, 148–49; lunar, 172
siderophile elements, 149–50
Slipher, Vesto, 39, 63
Smith, Bradford, 125
snow line, 130
snowball Earth episodes, 202–3
Soddy, Frederick, 61–62
solar nebula, 128, 129–33, 135–36, 137, 143, 144, 151, 152, 213, 214, 226, 230, 244, 256, 258, 261, 265, 266
spectrum, analysis of, 13
spiral nebulae, 52, 53, 63–64
Spitzer, Lyman, 54
s-process, 104
Stardust mission, 17, 245, 255
stars: formation of, 109–15; nuclear processes in, 98–106; oldest known, 74
stellar evolution, 72
Stern, Alan, 257
stromatolites, 191
subduction, 188
Suess, Hans, 95, 99
Sun: age of, 59–60, 71–73; birth environment of, 111, 119–21; composition of, 95, 96; energy source of, 3, 50, 59, 60, 61, 71, 100–104; evolution of, 118, 274–75; formation of, 114–15; physical properties of, 3–4; protoplanetary disk of, 121
supernova 1987A, 106–7
supernovae, 70, 73, 105–7; in Sun's neighborhood, 120

T Tauri stars, 118, 128
Taurus-Auriga star forming region, 113–14, 115, 118
technetium, 93–94, 104
Tempel 1, comet, 255
terrestrial planets, 5, 147; natural satellites (moons) of, 5; physical properties of, 148. *See also* Earth; Mars; Mercury; Venus
Terrile, Richard, 125
Theia, 179–80
Themis, 8
Thomson, William, 59–60
Titan, 11, 41, 218
Titius, Johann Daniel, 32
Titius-Bode law, 32–33, 34
Tombaugh, Clyde, 39–40, 246
transition disks, 138
trans-Neptunian objects, 8, 42, 248–49, 251–54; colors of, 253; densities of, 254
Trapezium cluster, 112, 119
tree of life, 192–93
triple alpha process, 103
Triton: discovery of, 39; fate of, 274; geysers on, 41; orbit of, 50, 218–19, 221, 256; origin of, 270; surface of, 257
Trojan asteroids, 7, 42, 259, 260, 261
Trujillo, Chad, 251
Tsiganis, Kelomenis, 259
turbulence: in protoplanetary disk, 136; in solar nebula, 130, 134, 135, 139
turbulent concentration, 136–37

uniformitarianism, 59, 60
universe: age of, 63, 64, 73, 74; expansion of, 64
Uranus: atmosphere of, 208; core of, 213; discovery of, 33; moons of, 50, 218; naming

of, 33; obliquity of, 216; orbit of, 37, 39, 40; rings of, 222. *See also* giant planets
Urey, Harold, 95, 99, 194
Ussher, James, 44

Valles Marineris, 161
Vega, 125, 127
Venera 9 mission, 141
Venus: atmosphere of, 158, 160, 265; compared to Earth, 140–41; formation of, 158; greenhouse effect on, 158; lithosphere of, 159; loss of water from, 158–59; rotation of, 160–61; space missions to, 11, 140–41; surface of, 159–60; transit of, 19, 20, 21
Vesta, 235, 270; crust of, 228; discovery of, 35; family, 233; and HED meteorites, 84, 90, 226; size of, 225. *See also* Rheasilvia
Viking missions, 41
Voyager missions, 41, 217, 222, 257

Walcott, Charles, 60
Walker, James, 200
Wallace, Alfred Russel, 44
Walsh, Kevin, 161
Ward, William, 136, 178
Wegner, Alfred, 189
Weizsäcker, Carl Friedrich von, 54
Wetherill, George, 230
Whipple, Fred, 243–44
white dwarf star, 104, 106
Wild 2, comet, 17, 131, 245, 255
Wilkinson Microwave Anisotropy Probe (WMAP) mission, 74
Wisdom, Jack, 78
Wolszczan, Alexander, 15

Yarkovsky effect, 79, 80, 232
Yarkovsky, Ivan, 79

Zach, Franz Xaver von, 34, 248
zircon crystals, 70, 189, 190